A PLANET IN PERIL

A PLANET IN PERIL

Global Warming and the Race to Find a Solution

Prem Shankar Jha

Rupa . Co

Published 2009 by

Rupa . Co

7/16, Ansari Road, Daryaganj
New Delhi 110 002

Sales Centres:

Allahabad Bengalooru Chandigarh
Chennai Hyderabad Jaipur Kathmandu
Kolkata Mumbai

Typeset in Adobe Caslon Pro by
Mindways Design
1410 Chiranjiv Tower,
43 Nehru Place
New Delhi 110 019

Printed in India by
Rekha Printers Pvt. Ltd.
A-102/1 Okhla Industrial Area, Phase-II
New Delhi-110 020

Contents

Preface

This book falls into three parts. The first reviews the growth of awareness among climate scientists and policymakers of the real threat that Global Warming poses. This is the threat of Abrupt Climate Change.

The second warns that the pre-Copenhagen negotiations—the world's first serious effort towards collective action—have wandered on to the wrong track: that of seeking consensus through confrontation instead of cooperation. The talks have become increasingly difficult because the negotiators have reversed the order in which issues need to be tackled: Instead of first agreeing on the most viable technology path to a low-carbon world and then calculating what shifting to it will entail, the negotiators have done the opposite. They have tried to set targets for reducing carbon dioxide (CO_2) emissions, without discussing how they will do so. They have bickered over who will finance the reduction of emissions, and how, without deciding precisely what they intend to finance and what its cost will be. They disagree on who, and to what extent, they will monitor each others' progress, without a clear idea of why this is necessary. This is like building a house without an architect's plan.

So the discussions are therefore taking place around a kind of Black Hole. A huge number of ideas, proposals and options are being thrown into it, but nothing is coming out. The result is a rising impatience that is being reflected in the international media. Since there is no consensus on the technologies to be developed and exploited, the industrialised countries have no idea what they will be committing themselves to if they agree to part-finance the energy shift in the developing countries. They are therefore keen to minimise their obligations to the poor. The developing countries are aware of this; so they distrust any promises of compensation that the rich are prepared to make.

The third, and most important, part of the book is therefore devoted to filling the hole. It proposes a family of technologies and a set of associated policies that can usher in a low-carbon and renewable-energy economy in the four decades that by common consensus, is all the time left for the world to avert catastrophe. These technologies share four prerequisites for reaching the goal, of reducing the emission of greenhouse gases by half by 2050, that the international community has set. First, most of their component elements of these technologies are already technically proven and are in commercial use. They can therefore be harnessed to start the shift out of fossil fuels straightaway.

Second, their economic viability has been assessed with a fair degree of precision and numerous studies have concurred that most of them are either viable already or close to being so. They will therefore need only a limited amount of subsidisation to make them attractive to investors.

Third, they can be inducted without disturbing the existing patterns of production and division of labour, whether in agriculture or industry. Thus the cost of the energy shift in terms of investments wasted, towns turned derelict, and human skills rendered redundant, will be close to nil.

Last, apart from stabilising, and then reducing, the concentration of carbon dioxide in the atmosphere, these technologies will confer a host of other economic and social benefits upon society.

This book is the product of research that I began twenty-seven years ago in the aftermath of the twelve-fold rise in oil prices that occurred between 1973 and 1979. My limited purpose then was to look for biomass-based alternatives to gasoline and diesel, in order to reduce the crippling burden of oil imports on India. Since there was one form of biomass that was not in short supply in India at that time—municipal solid waste—I proposed the gasification of MSW to produce not electricity, which made little economic sense, but methanol as a fuel additive, which promised save gasoline and diesel and be profitable at the same time. Biomass gasification technologies were then in their infancy. So, based on a directive from then Prime Minister, Mrs Indira Gandhi, the Maharashtra government set up a study team in November 1982 to appraise available technologies for the gasification of municipal solid waste. I was made a member and *rapporteur* of that team. The report submitted by the team in March 1983 contained techno-economic appraisals of seven processes, and recommended two for further study and development at the pilot plant scale (It also recommended drying and pelletising the solid waste to increase the energy content of the feedstock and reduce the capital cost of the plant).

However, the project fell foul of a strong lobby in Mumbai and Delhi that favoured importing waste incineration systems to produce electricity. The lobby prevailed, and one incineration plant with a capacity of 300 tonnes per day was set up in Timarpur, a north-eastern suburb of Delhi in 1983. Today, 26 years later, the plant has worked for a total of five days: three days during a demonstration set up for Mrs Indira Gandhi, and two more days during a demonstration for her successor Rajiv Gandhi. On both occasions the plant managers had to add fuel oil to the MSW in order to make it burn. I fear that most of the carbon capture and storage plants being set up in Europe and the US today will end up the same way.

In 1986 oil prices crashed, and the world lost interest in renewable energy. By the time oil prices once again crossed the $40-barrier it was 2004, and almost two decades of time that the world does not have, was lost. This book is an attempt to make up for the lost time.

Chapter 1

Business as Usual

In June 2008, the G8—comprising the world's main industrialised nations—arrived at a momentous decision: Carbon dioxide emissions had to be lowered by fifty percent in absolute terms by 2050 or, it warned, the future of the human race was bleak. While making for interesting headlines, the pronouncement was riddled with ambiguity for it did not spell out the base year for setting the targets for emission reduction. It was, nonetheless, a momentous occasion, for it marked the return of the US from the wilderness into which the administration of George W. Bush had taken it when he pulled the US out of the Kyoto Protocol. It was thus not surprising that the Summit attracted a large amount of media attention.

It was not the first time that a public pronouncement from world leaders on the issue of global warming had evoked a flurry of interest. In 2005, Former US Vice-President Al Gore's two-hour video documentary on climate change, titled, 'An Inconvenient Truth', shook up the entire world. It inspired the then UK Prime Minister, Tony Blair to make global warming one of two central issues to be discussed at the G8 meeting in Scotland that year. But the entire meeting was swept out of the public mind by the London terrorist bombings that preceded it. Two years later, by the time that Gore and the Inter-Governmental Panel on Climate Change were awarded the Nobel Peace Prize it seemed that the world had forgotten their message.

In much the same way, interest in the issue faded out soon after the Hokkaido announcement as other, more immediate, crises captured world headlines. When two months later, British Prime Minister Gordon Brown made it clear that to reach a fifty percent reduction by 2050 the industrialised countries needed to reduce CO_2 emissions by eighty percent, he provoked transient interest, but still no alarm. Few outside the climatologists' community, even in Britain, bothered to ask why he had done so. The level of public awareness rose again after the Copenhagen Climate Change Science Congress in March 2009, when 2,500 scientists issued a collective warning that the world would simply have to bring down CO_2 emissions by more than fifty percent if it hoped to avoid 'dangerous climate change'. Since then it has been sustained by the growing acrimony of the pre-conference negotiations on one hand, and renewed attempts to debunk the very idea that human beings can contribute more than a negligible amount to global warming, on the other.

The two have very largely cancelled each other out. As a result, with no serious public pressure upon them, lawmakers have, preferred not to make any departure from their 'business as usual' behaviour practices. This was highlighted by the debate that occurred in the US House of Representatives almost exactly a year after Hokkaido, on the Waxman-Markey Clean Air Bill that sought to reduce carbon dioxide emissions in the

US. Most of the 212 Congress members who voted against the bill (219 voted for it) did so because they simply did not believe, or chose not to believe, that global warming posed a serious threat to humanity. The defining moment in the debate came when Paul Broun, Republican Congressman from Georgia, argued that global warming was 'a hoax perpetrated out of the scientific community'. Broun did not stop to think that for such a gigantic hoax to be pulled off, it would have required 'a vast cabal consisting of thousands of scientists—a cabal so powerful that it has managed to create false records on everything from global temperatures to Arctic sea ice'.[1] Worst of all, Broun was actually cheered by his fellow Republicans in Congress, not one of whom voted for the bill.[2]

But not thinking about a threat will not make it go away. If people have been choosing to ignore the threat that global warming poses, it is in large part because, not knowing how to avert it, they have continued to put their blind faith in Science and the Market to find them a solution. After all have these twin gods not come to humanity's rescue time and again in the last two hundred years?

The Market has been quick to capitalise upon their faith. In the past two decades manufacturers have flooded the market with 'green' goods and processes that claim to be energy-saving and environment-friendly. Several of them genuinely are. But overall, however, they have made no dent in the exponential rise of CO_2 emissions. Some of these innovations have actually had the opposite effect: instead of reducing the threat from global warming, they have increased it by dulling the edge of the public's anxiety and weakening its will to act. At least three recent 'breakthroughs' illustrate how Science, in the service of the Market, has managed to do such disservice.

In June 2008 the Japanese car company, Honda, unveiled the world's first commercially available, totally environment-friendly car. Propelled by the long-awaited hydrogen fuel cell, it is a family saloon capable of accelerating from zero to a hundred kilometres per hour in 9.8 seconds, and of running more than 400 kms without needing a recharge of its fuel cells. Honda is leasing it to selected customers at a mere $600 a month. A shade expensive—about the lease price of a Porsche—but still just about affordable. Should the world then breathe a sigh of relief? Has technology provided an answer to one of the principal causes of global warming? A closer examination of Honda's plans shows that its hydrogen fuelled car has relatively little to do with global warming.

The cost of the car is not around $100,000, which is what the $600 monthly rent would lead us to expect, but 'several hundred thousand dollars' (probably around $300,000). Honda is producing only twenty-six of these cars and 'selling' them at a hugely subsidised price to a few, very high profile, celebrities in order to gain maximum exposure for it. The company expects to bring down the cost to $100,000, but this will take another ten years, if all goes well. By then the rising cost of gasoline may make it an attractive, not to say novel, alternative to the Porsche or Ferrari. But even ten years hence the car will be little more than an expensive toy. It therefore has as much to do with reducing global warming as a Cartier necklace or a Ferragamo handbag. Honda knows this. It has therefore developed a marketing strategy that will ensure that the fuel-cell driven car will have a ready market among the super-rich of the world when its cost-saving technologies are finally in place.

Two other much touted technologies also flatter only to deceive. The first is the much discussed burying of

CO_2, generated by coal and oil burning power plants, deep underground. The second is the partial replacement of gasoline with ethanol.

Capturing CO_2 from power station emissions is easy. Storing it underground on any significant scale is not only difficult but expensive as well. At a 120-MW zero-emission, coal-based, demonstration plant being set up by a German company Vattenfall Europe A.G. will soon begin to capture nine tonnes of CO_2 per hour, or 85,000 tonnes a year. The promoters, who have been awarded a large subsidy, intend to store it deeply and securely underground and have ruled out the possibility of its escaping.

The example set by Vattenfall has taken a world that craves 'quick fixes' by storm. As of April 2009, various governments had approved 238 Carbon-Capture-and-Storage (CCS) projects, saying they were willing to subsidise them. Three-fifths of these are at the research, pilot or demonstration stage, but no fewer than a hundred are commercial-scale plants[3]. But how secure is the Earth's crust itself as a storage place? Since earthquakes have been known to break clathrates and release vast amounts of CO_2 into the atmosphere, storing ever larger quantities of CO_2 under high pressure underground would be like hanging a sword of Damocles over humanity's head, and watching it grow larger with every passing year.

EU policy makers have been struggling to reassure the communities under whom they intend to store the CO_2 that it is perfectly safe to do so. But they are facing growing resistance to the very concept[4].

The quantities that will have to be stored, moreover, are mind-boggling. The annual release of CO_2 from power plants burning fossil fuels exceeds eleven billion tonnes. The British scientist, James Lovelock, has estimated that, if frozen, the CO_2 we produce would create a mountain a mile high and twelve miles in circumference every year.

Carbon capture will also be prohibitively expensive.Among the approved projects, those with known costs were estimated to require an investment of $44.5 billion to sequester sixty million tonnes of CO_2 per year. This amounts to $741 per tonne of CO_2. To capture all of the energy-related CO_2 emissions will therefore require an investment of $22.25 trillion, and all of this would have to be justified on the grounds of its social benefits, for it would not yield even one additional dollar of revenue to the power plant.

Estimates made at the National Energy Technology Laboratory and Massachusetts Institute of Technology (MIT) in the US, suggest that carbon capture and storage will add between seventy-five and ninety percent to the capital cost of older plants and sixty to eighty-five percent to the cost of newer plants working at higher, supercritical, temperatures. According to the fourth IPCC report the cost of electricity from a coal-based carbon capture plant will be double the cost of power from a natural-gas-based combined cycle plant. But this is only an estimate, as no such commercial-scale plant has been built so far. Finally, undersea storage, the third option, is poorly understood, riddled with environmental side effects, and will in any case only delay the eventual release of the CO_2 into the atmosphere.

Perhaps the mosr disastrous experiment tried so far has been the replacement of gasoline with ethanol. Launched by President George Bush's offer of a 51 cent a gallon subsidy on ethanol in his State of the Union address in 2006, in three short years it has led to the diversion of between twenty and thirty percent of the American corn and soya bean crop from feeding human beings and cattle to the production of ethanol. This triggered a sharp rise in price of corn, a staple food, in Mexico and the conversion of large tracts of Brazilian ranchland into soya bean fields and an acceleration of deforestation in the Amazon to create new grazing lands. Even if the technology had not had these undesirable side effects, it would never have met more than a small fraction of the transport fuel needs of the world. But as the sixty-five-percent rise in the global price of food grains between the beginning of 2006 and the spring of 2008 showed, the side effects are so severe that even as a partial replacement, it is a complete non-starter. For it is now apparent that even a minimal diversion of food crops to the manufacture of ethanol can push millions of people in the poor countries over the edge of hunger into starvation.[5]

But the climate sceptics have been losing ground. Notwithstanding the huge distractions created by a surfeit of information, public attention has focussed more and more on the threats, both real and imagined, from global warming. President George Bush's dramatic return to the Kyoto process at Hokkaido in 2008, and the passage of the Waxman-Markey Bill by the US House of Representatives, shows that the believers in leaving it to Science and the Market, are losing ground. Notwithstanding the presence of lingering doubts over the extent to which human activity is responsible for global warming responsible people from all walks of life are veering around to the view that they cannot gamble on the climate scientists' predictions proving wrong. For what is at stake is not simply millions of species of flora and fauna, and thousands of atolls which are projected to disappear under the rising seas, but civilisation itself.

Chapter 2

'Abrupt' Climate Change

Till the end of the 1990s, environmentalists had believed that the changes being wrought by global warming—a rise in the mean temperature and sea level, and an increased frequency of floods and droughts, would take place gradually. This would give human beings time to adapt as they, and millions of other living species, have adapted to previous changes. In its most plausible form this belief was expressed by the popular writer, Michael Crichton, in his 2004 book, *State of Fear*. The book sold 1.5 million copies in its first print run. Crichton ended it with a twenty-page dissertation on how various species of fauna and flora had vanished since the beginning of life on earth, only to be replaced by others. The human species had not only adapted and survived, but thrived. This best-selling piece of pseudo-science immensely reinforced the sceptic views on climate change, and became the bible of the neo-conservatives who dominated the US administration in the Bush years.

The irony is that Crichton's book came out precisely at the time when the changes being caused by global warming had become impossible to ignore. It was the year when the Arctic

Ocean melted and clear water appeared in it in summer for the first time; chunks of the Antarctic began to break off and fall into the sea; and 3,000 people died of a heat wave in Europe that took temperatures in London up to 40 degrees Celsius.

In 2001, when the *New York Times* sounded the alarm bells when it featured photographs of Tanzania's Mount Kilimanjaro, which showed that at its present rate of shrinkage, there would be no ice left on Africa's highest peak in only fifteen years[6], it aroused a storm of controversy. Various scientists, writing in *Nature* and the *American Scientist,* said that while the shrinkage of glaciers nearer to the Poles could indeed be caused by global warming, Kilimanjaro, being an equatorial, glacier, was governed by different rules[7].

Glaciers are in retreat across the world. The South Cash glacier in the Rocky mountains has lost half of its ice between 1928 and 2001. New Zealand's Southern Alps has lost eleven

The Upsala glacier in the Southern Andes, in 1928 and 2004. Photo provided by Greenpeace and published in Time Magazine, New York. March 26, 2006. "Global Warming Heats Up".

percent of its ice in the past thirty years[8]. Glaciers are also in retreat in the Andes and the Himalayas. Even in the case of Kilimanjaro, the argument that the loss reflects a decline in precipitation caused by deforestation, is not sufficient to explain the sheer magnitude of the shrinkage of its ice cap—eighty-two percent between 1912 and 2001.

Nor can it explain the sudden acceleration of the shrinkage in the last two decades, for fully one-third of the ice has disappeared after 1989. All across the world, arid sub-tropical zones have expanded an average of 250 miles (400 kms) towards the Poles.

The icecap on Kilimanjaro, at the border of Kenya and Tanzania near the equator, seen in 1912.
Kilimanjaro photos courtesy Lonnie Thompson, Ohio State University.

In the last two decades, weather patterns have changed dramatically. During 1988, the year when climate scientist James Hansen first warned the US senate that global warming had already arrived, was the hottest year in recorded human history. But 14 of the ensuing 20 years have been even hotter.

Kilimanjaro's ice cap lost 82 percent of its area by 2000. Courtesy Why files. Posted 23 Feb 2001.

The sustained heat is causing unmistakable changes in the earth's geography. The Arctic Ocean is no longer frozen. At its present pace of melting, there will be no ice left in summer in only five years. Large chunks of the Antarctic ice fields—the size of small countries—are breaking off and falling into the sea. In 2006 the IPCC (Intergovernmental Panel on Climate Change) reported that the temperature of the top layer of the Permafrost in the Arctic region had risen by three degrees Celsius since the 1980s. Just two years later, in the summer of 2008 in Alaska, the permafrost began to melt for the first time in human history. As a result, roads are cracking up and lakes are disappearing as the water dribbles away into the earth's crust. Alaska, paradoxically, faced a water shortage in 2008.

Although statistical evidence on the number of storms, tornadoes, and hurricanes, taking place every year shows no distinct trend, weather conditions seem to have become more

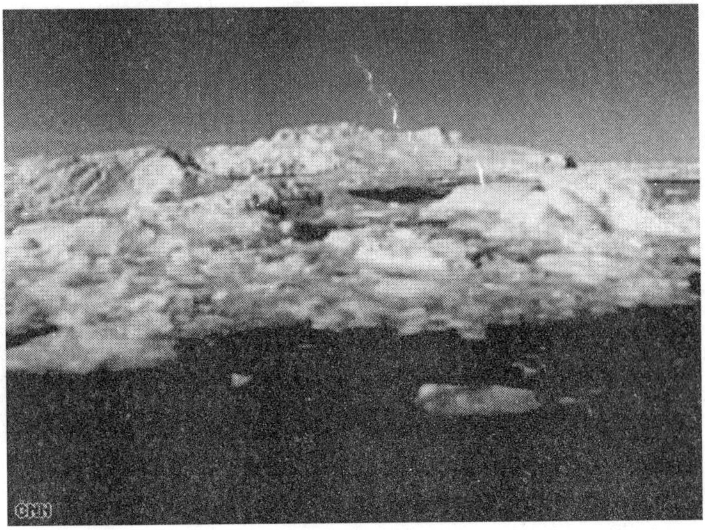

Photo of the Arctic in summer.

extreme and more unpredictable with every passing year. The year 2004 saw the onslaught of a heat wave across Europe. In 2005, hurricanes of unprecedented ferocity pummelled different parts of the world. Among these were Hurricane Larry, in north-east Australia which built up winds of 180 miles an hour, and Hurricane Katrina that destroyed much of New Orleans, in the United States. Katrina was one of four hurricanes of exceptional ferocity to hit the US that summer.

In 2006, a prolonged drought made the annual burning of forests in Indonesia go out of control, and blanket much of Asia in smog for months. Heaven alone knows how much additional CO_2 was released into the atmosphere by that catastrophic event. In the beginning of 2008, prolonged drought made the rice crop fail in Australia, even while much of the rest of the world was experiencing unusual, if not unprecedented, rains, snowfall, and cold weather. In May and June of the same year the US and China found themselves reeling under severe floods caused by torrential downpours. In 2008 the monsoons arrived three weeks early in India and were unusually heavy. The very next year they arrived several weeks late and were still deficient by a quarter in the middle of September, when the rainy season was drawing to a close. In 2008 the Philippines was devastated by an unseasonal typhoon, and in 2009 the same thing happened to South Korea. The long delay in the Indian monsoon in 2009 was caused, in part, by three unseasonal and wholly unexpected cyclones over north-eastern India.

The Point of No Return

The sheer speed at which everything is happening has given credence to a theory that, till only a decade ago, had been dismissed as the croaking of a Cassandra. This is the theory of the 'Tipping Point'. Paleoclimatic research, based on the analysis of core samples from the Greenland ice cap, has been suggesting that climate change does not have to be gradual. On a number of occasions there have been changes in the mean atmospheric temperature of as much as sixteen degrees Celsius within as little as a single decade. The end of the ice age, 15,000 years ago, was one such event. The descent into another mini-ice age, known as the Younger Dryas period, 12,800 years ago and ascent out of it 11,600 years ago, was equally sudden. There was another smaller 'cold event' 8,200 years ago that showed a similar pattern.

Despite the gathering momentum of climate change around us, even the scientists who know most about global warming—those who have devoted their entire lives to charting the rise of sea and land temperatures, the concentration of greenhouse gases in the atmosphere, and the effect this is having on flora and fauna—are strangely reticent when it comes to putting into words the conclusion to which their accumulated data are driving them. This is that we are now perilously close to a point of no return. Were we to cross it, the future of the planet—and of human civilisation—would slip out of our control. Even bringing down future GHG emissions to zero will not save the earth, or most of the species that currently live on it because the earth will be trapped in a self-reinforcing cycle of heating. Scientists even have a name for the point, or points, at which this is likely to happen: the 'Tipping Point'.

The idea of a Tipping Point has been around for more than three decades. One of the earliest, formulations, and certainly the most compelling, has been by the celebrated scientist, James Lovelock. Lovelock identified two tipping points: the melting of the arctic ice, the Greenland ice cap, and parts of the west Antarctic ice sheet, and a warming of the oceans that leads to the death of most of the algae in it. The former will progressively reduce the area of the earth's surface which reflect the sun's rays back into space without allowing it to heat the earth. The latter will eliminate the single most important carbon sink on the planet, for sea algae trap two-thirds of the CO_2 that is naturally created and, even today, sequesters forty percent of the thirty billion tonnes of CO_2 being spewed into the air by human beings. Both these processes increase the rate of global warming, the first by exposing more and more of the surface of the sea where the water can absorb the sun's heat, and the second by accelerating the accumulation of CO_2 in the atmosphere.

Where will these processes end, and what kind of planet will they leave behind? Most scientists are unwilling to give an explicit answer because there are still too many influences at work that they do not fully understand. But there is a growing consensus that these processes will lead to a sudden sharp rise in temperatures, a rise in sea level, the inundation of large parts of the currently inhabited parts of the planet, and desertification of several of the warmer parts of the globe. The suddenness of this change was described vividly by Jeffrey Kluger in *Time* magazine in 2006:

> Global climate systems are booby-trapped with tipping points and feedback loops, thresholds past which the slow creep of environmental decay gives way to sudden and self-perpetuating collapse. Pump enough CO_2 into the sky, and that last part per million of greenhouse gas behaves like the 212th degree Fahrenheit that turns a pot of hot water into a plume of billowing steam. Melt enough Greenland ice, and you reach the point at which you're not simply dripping melt-water into the sea but dumping whole glaciers. By one recent measure, several Greenland ice sheets have doubled their rate of slide, and just last week (2006) the journal *Science* published a study suggesting that by the end of the century, the world could be locked in to an eventual rise in sea levels of as much as 20 ft. Nature, it seems, has finally got a bellyful of us.[9]

Lovelock's analysis of the geo-climatic history of the earth suggests that it could lead eventually to another ice age of indeterminate depth and duration. This can be deduced from his model of earth as 'Gaia'. In 1979 Lovelock pointed out that the earth was not a semi-molten piece of inanimate matter, cast off from the sun that just happened to go into orbit around it at the right distance for a crust to form and trigger the sequence of events that have led to the creation of sentient life. On the contrary, living creatures evolved *with* the evolution of the earth's eco-system and both shaped and were shaped by it. Today, three billion years after the birth of life on the planet, its fauna and flora, its atmosphere, its forests, its ice and snow cover, its forests and its sea and land temperatures, are not governed by separate systems but are part of a single interdependent system whose stability depends upon its not being disturbed. Every disturbance of one of its elements causes the rest of the components to adjust in order to restore equilibrium. In that sense the earth as a whole behaves very much like a living organism that fights back to repel invaders. If the disturbance of the equilibrium is sufficiently powerful and sustained, at a certain point the earth will shift all of its parameters to establish a new equilibrium. The change from the old to the new equilibrium will not be gradual but is likely to take place within as short a span of time as a few decades.

The key driver of change is the energy the earth receives from the sun which is getting hotter with the passage of time. Three billion years ago the earth received about a quarter less energy from it than it does today and was a rather dark place. Its atmosphere contained thirty times as much CO_2 as it does today. There was very little oxygen. The earliest living organisms developed in the absence of oxygen, and many are still around as anaerobic bacteria that only have to be exposed to oxygen to die.

In that early earth the decay, after death, of these organisms created methane, which is twenty-four times as potent as a greenhouse gas as CO_2. The resulting blanket of GHGs created a huge greenhouse effect, but it was one that worked in reverse direction from today, for it kept the heat contained within the earth, and that generated by the decay of dead matter, from escaping into space.

But as the sun aged and grew hotter, the need for conserving the earth's own heat declined. Temperatures rose on the earth's surface and the thick mantle of CO_2 and methane became a hindrance. Aerobic—meaning oxygen-breathing living matter began to multiply. The miracle, when it occurred, was photosynthesis. This steadily lowered the concentration of CO_2 and raised that of oxygen. It began with the growth of Algae in the oceans and, as the oxygen content of the atmosphere rose, spread rapidly to the land. Lovelock christened this 'active', almost sentient, earth 'Gaia'.

In his estimation the earth was healthiest about a billion years ago. At that point the sun's heat was just sufficient to make it unnecessary to either conserve or disperse any of it. But as the sun continued to grow hotter, the earth entered a period in which it had increasingly to manufacture the conditions in which the verdant and abundant life of that period could continue to flourish. It did so by generating ice ages that cooled the earth down. These had several effects: they cooled the earth by increasing the size of the ice caps; they cooled the oceans and revived its rich plant life, thereby increasing its capacity to sequester carbon. They reduced the land area that could sustain life, but also exposed large amounts of fresh new land for the spread of flora and fauna by lowering the sea level by as much as 120 metres.

The proof that the ice ages restored the earth's health is to be found in its very low concentration of CO_2 in the atmosphere—180 parts per million (ppm). This was fully 100 ppm below the level that prevailed from the end of the last ice age till roughly 1850 AD and 210 ppm below the level attained today. As Lovelock notes, it needed a lot of vegetation to pump the CO_2 concentration down so low.

But as the aeons have passed and the sun has continued to grow hotter, even 'Gaia's' capacity to regulate herself has begun to approach its limit. Lovelock estimates that even without human interference the earth had about a hundred million years to go before it becomes a dead planet. Human intervention has come at a time, therefore, when its capacity to adjust is close to being exhausted.

Based on his studies and the rapidly accumulating evidence collected by climate scientists all over the world, Lovelock believes that the earth is already in the throes of another abrupt climate change. In 1994 he and geochemist, Lee Kump, published an article in *Nature*, which attempted to understand the behaviour of the earth with the help of an elaborate computer model that included just about every variable they could think of. The variables included the behaviour of ice caps, tropical forests, and algae, and small periodic variations in the heat of the sun. In the model, the concentration of CO_2 was trebled over a period of 20,000 years from pre-industrial levels, a rate far slower than the pace of accumulation today. They came up with surprisingly unambiguous answers. As the concentration approached 500 ppm, amounting to a temperature increase of about three degrees Celsius, two things happened almost at the same time: The Greenland ice cap became unstable, and the changes in land vegetation and ocean algae became more pronounced, suggesting that a self reinforcing mechanism had been set in motion. At 500 ppm the melting became irreversible and the capacity of the algae to absorb CO_2 nose-dived. In their model, within decades, the mean temperature rose by sixteen to twenty-four degrees Celsius!

In Lovelock's opinion the situation today is similar to that which preceded the sudden onset of an extremely hot period fifty-five million years ago, called the Eocene period, when the earth suddenly warmed by eight degrees Celsius in the temperate regions and five degrees in the tropics. The hot period this set off lasted for 200,000 years. Geological studies by Professor Harry Elderfield at Cambridge University suggest that it was triggered by a sudden release of between 0.3 and three trillion tonnes (terra-tonnes) of fossil carbon into the atmosphere. Lovelock concludes by warning his readers that human beings have added half a trillion tonnes of carbon to the earth's atmosphere in the past century, and the rate of accumulation is rising exponentially. The earth is therefore within the range of the event that triggered the onset of the Eocene hot period.[10]

Chapter 3

A Conspiracy of Silence?

If the stakes are indeed so high, why is there almost no one (other than former US Vice-President Al Gore) warning the world of the price of inaction? Why is no one telling them that the quest for prosperity, if based upon the continued burning of fossil fuels, will lead to a violent death if not of our generation then, very possibly, of our grandchildren? The IPCC goes to almost absurd lengths to hedge all but the most unequivocal of its conclusions, claiming an insufficiency of field data. But how much data is 'sufficient'? When does the quest for more data become an excuse for avoiding conclusions that one would rather not face? A detailed analysis of the Stern and IPCC reports suggests a simpler answer. The scientific community, and with them the governments of the world, are unwilling to acknowledge the full magnitude of the threat that hangs over them because they do not know how to meet it.

Climate scientists are aware that Abrupt Climate Change is a real possibility. But lacking solutions, they are reluctant to make predictions that will generate worldwide panic, and trigger precisely the kind of dog-eat-dog scramble for safety that can bring human civilisation to a tragic end.

Both the Stern and IPCC reports have charted the rise in air and sea temperatures in the past hundred years at different latitudes, over land and over the sea, correlated it to the rate of increase in the concentration of GHGs and made projections of the increase in average temperatures that will take place in the next hundred years if the concentration is allowed to rise unchecked. Both have also predicted that as the CO_2 concentration increases and the earth grows warmer, the sea level will rise, glaciers will shrink, the area of the earth's surface under ice and snow will decline, fertile coastal plains will be submerged causing hundreds of millions of people to move inland, and fresh water will become scarce, especially in the warmer regions nearer the equator.

Both agree that the weather will become more extreme and more unpredictable, and that there will be more floods and longer droughts. Overall, this will reduce the earth's capacity to produce food and, once again, it is the poorer countries that will suffer the most.

But both have very little to say about what is perhaps the most dangerous development in recent years, the decline in the capacity of the oceans to absorb CO_2 from the air. While they currently absorb around half of human CO_2 emissions, empirical observations reveal that they are rapidly losing the ability to do so. A study released in May 2007 found that the Southern Ocean has reached its saturation point, and that its ability to absorb CO_2 could no longer offset any part of the increase in emissions. [11]. Similarly, CO_2 absorption by the North Atlantic has been cut in half over the past decade. A 'positive feedback' wherein the oceans take up less CO_2; the CO_2 concentration in the atmosphere rises faster; global warming accelerates, and this further reduces the ocean's capacity to absorb CO_2, may already be in place. Indeed Lovelock's grim prediction is perilously close to coming true.

With no solutions to offer, both the Stern and IPCC reports implicitly or explicitly reject the likelihood of abrupt climate change.[12] They do this by charting a variety of curves that reflect the CO_2 levels that will be reached by the end of the century under various scenarios, The implicit assumption is that the structure of human society will still remain fundamentally unchanged even when it reaches levels of CO_2 concentration of 750 ppm or higher. Adjustment will be extremely painful but will still be possible. The two reports therefore rule out the possibility that the earth could start an adjustment process of its own that takes away humanity's control over its future, by refusing to discuss it.

But while the IPCC has fought shy of making apocalyptic predictions it has scrupulously presented the data for people to draw their own onclusions. These do not rule out the possibility that the regulators of global climate that have kept it stable for a billion years may already have begun to fail. It has established that although the mean air temperature rose by between 0.7 and 0.8 degrees Celsius between 1906 and 2005, the rate of increase in the average land temperature has been twice as high in the second 50 years, as in the entire period hundred years from 1906 to 2005. A simple calculation shows that this means that two thirds of the warming has taken place after 1956. The actual rise may have been even sharper. Between 1940 and 1970 the earth went through a small cooling period which actually brought down the mean global temperature inspite of a steady rise in CO2 concentrations. This effect ended in the beginning of the seventies. In the ensuing thirty-four years the mean air temperature over land has risen by two-thirds of a degree Celsius.

But global averages tell us very little. What we need to know is by how much the temperature has risen over the Greenland ice sheet and the Arctic ice cap. The findings reported by the IPCC report are disturbing enough, noting that the rise in temperatures has been twice as great in these regions, Northern Russia and China, and still greater—up to 3.5 degrees—in Alaska and parts of northern Canada. The observed effects are bearing out the accuracy of Lovelock and Kump's model. For as the rise in temperature in Greenland has approached, and possibly crossed, the two-degree mark, the rate of melting of the arctic ocean's floating ice and the Greenland ice sheet has picked up sharply. The IPCC's data show that the contribution made to rising sea levels by the melting of the arctic ice has doubled, and that of the Greenland ice cap quadrupled between 1993 and 2004.[13]

The IPCC has predicted that mean temperatures will rise in the next two decades by 0.2 degrees per decade.[14] But the two to four times higher increase that has taken place in the arctic and polar region during the last thirty years indicates that it could rise by 0.4 to 0.8 degrees per decade. By 2025, therefore, the mean temperature could increase by 0.8 to 1.6 degrees in the Greenland, Arctic, and Siberian regions, and by even more in Alaska and northern Canada. This would take the average rise in atmospheric temperature in these northern regions to three degrees or more since 1970. The earth is thus set to cross the 2.7-degree threshold for the irreversible melting of the northern ice predicted by Jonathan Gregory, whose studies Lovelock has cited, in less than 20 years. This will be sufficient to melt the Greenland ice cap and raise the sea level not by the 0.5 to 1.7 metres that policy makers are hoping to limit it to, but by seven metres!

Many other scientists have predicted even larger increases in temperature. The Stern Review has pointed out that if the CO_2 concentration rises to 550 ppm then there is a seventy-seven to ninety-nine percent chance (i.e., a near-certainty) that the average temperature will rise in the arctic and Greenland by two degrees Celsius. At the current rate of increase in CO_2 emissions the 550 ppm barrier will be breached by 2035. At that point, models of abrupt climate change will no longer be needed for we will already be in the midst of one.

The world thus has less than thirty years in which to avert catastrophe. One well-known scientist who has fully realises how little time the world has left, and has decided to speak openly—even at the risk of causing panic is James Hansen, the man who delivered the original warning to the US Senate on global warming twenty years ago. On 23 June 2008, he warned the Senate that the world had frittered away in the past twenty years doing nothing, or worse even, indulging in tokenism while the industrial lobby carried on with 'business as usual'. The Tipping Point, he warned, was now very near. Indeed it could already have been crossed, for the arctic ice was melting, increasing the area of clear water and therefore the area that could absorb, instead of reflect away, the sun's heat. As a result, even if there was no further addition to the CO_2 in the atmosphere the arctic would soon be free of ice. The goal that the world had set for itself, he concluded, of containing the rise in temperature to 2 degrees Celsius in the land areas over the rest of the current century, was not therefore a recipe for salvation but for global disaster. Far from allowing it to increase, the CO_2 concentration had to be brought down from 386 ppm to where it had been in the late '80s, i.e. 350 ppm. The world had, at most, two more decades to do so.

Otherwise he warned the National Press Club later that day, 'we are toast'.[15]

In quantitative terms what does this target actually require us to do? Hansen's warning provides a provisional answer. To do this we not only need to arrest the annual growth in CO_2 concentration of two parts per million but bring it down by 1.4 ppm every year. This does not sound like a great deal till we remember that the 2 ppm increase is occurring because the earth is able to absorb only half of the greenhouse gases (amounting to two-thirds of the CO_2) that humanity generates every year. A fifty percent reduction is necessary only to keep emissions from rising further. What is worse, the earth can absorb no more CO_2 but the consumption of energy is not standing still. Under a Business As Usual scenario CO_2 emissions will increase by another fifty-five percent to 46 billion tonnes by 2030[16] and somewhere close to 60 billion tones by 2050. This means that the world will have to avoid forty billion tonnes of emissions simply in order to bring CO_2 emissions into balance with the earth's capacity to absorb them. CO_2 Halving CO_2 emissions by 2050 may not therefore suffice.

Is it even remotely possible to make such a drastic shift in the next forty years? Projections made by Exxon-Mobil suggest that it is not. These show that if the rate of growth of ethanol (produced from cellulosic sources other than food grains) is doubled, if the rate of efficiency growth in new cars is also doubled, if half the addition to coal-fired power is replaced by nuclear plants, and if all coal-fired plants that are more than forty years old are retired, CO_2 emissions will still rise to thirty-two billion tonnes per annum by 2030. Bringing them down by even a small percentage, let alone by two-thirds, will require measures that Exxon was not even willing to contemplate because it does not think they are feasible.

Global CO_2 Emissions

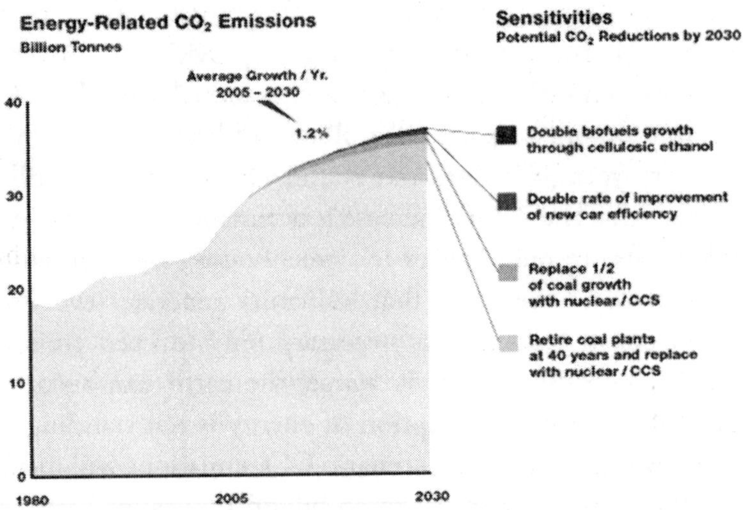

Figure 1: Global Co_2 Emissions

Source: Exxon Mobil Forecast: Global CO_2 emissions.
http://www.exxonmobil.com/Corporate/energy_outlook_globalco2.aspx

This is why the entire scientific community is playing a game of make-believe. No one—not the team that wrote the Stern Report, not the authors of the Fourth IPCC report, and not the leaders who made the Hokkaido commitment—really believes that the goal of halving emissions by 2050 can be achieved. This accounts for the deep pessimism that runs through both major reports.

The Stern Report devoted a whole chapter to mapping out alternative scenarios of what will happen if CO_2 accumulates at various rates, and the magnitude of the effort that will be needed to bring it down. It never leaves one in any doubt that the CO_2 level will have to be brought down. What it does not tell us is how to do it. The IPCC Report follows a similar path and has even less to suggest by way of a solution. The Exxon Mobil study is one of the few that has attempted to chart an energy transition path out of fossil fuels, and its conclusion is that all the methods it has outlined, put together will not even stabilize CO_2 emissions at their present level. Lurking behind these evasions is the fear that there is not much that anyone can do to stop the concentration of CO_2 from rising beyond 450 ppm, possibly reaching 550 parts per million before the middle of the century.

The private sector has not been so timid, and has rushed in where the scientists have feared to tread. Their *deus ex machina* is energy conservation. The McKinsey Global Institute (MGI), the research arm of the consultancy, thinks that energy efficiency could get the world halfway to the goal, espoused by many scientists, of keeping the concentration of greenhouse gases in the atmosphere below 550 parts per million.

MGI is particularly enthusiastic because it believes that unlike most other schemes to reduce emissions, a global energy-efficiency drive would be profitable at current prices and would therefore require no national or internationally coordinated interventions in the market. The measures it has in mind, all of which rely on existing technology, would earn an average return of seventeen percent and a minimum of ten percent. The IPCC leans towards the same view. It believes, for example, that profitable, efficiency-enhancing, investments would allow Pakistan to cut its emissions by almost a third, Greece by a quarter, and Britain by more than a fifth.

But as the data presented above, and Exxon Mobil's projections, show, the world is far beyond the point where greater energy efficiency alone will avert the impending catastrophe. This accounts for the apocalyptic tone that James Hansen adopted when he appeared before the US Senate on 23 June 2008, and the deep vein of pessimism that ran through his proposals to the US senate. These were:

1. To phase out the use of coal except where the carbon is captured and stored below ground. This, in his opinion, is the primary requirement for preventing further global warming.
2. To move to carbon-free energy sources promptly. In particular, pass a global ban on the substitution of transport fuels derived from oil with fuels derived by squeezing oil from coal, tar shale, or other fossil fuels.
3. To prosecute for crimes against humanity and against the planet, the CEO of any fossil fuel company that directly, or indirectly, blocks a transition to renewable energies, and advocates 'business as usual' policies for energy generation.
4. To impose a carbon tax on all fossil fuels at the point of extraction or import that reflects the social cost of using them.

These suggestions are unexceptionable, but easier proposed than implemented, especially in the two decades that are left before the world crosses the Tipping Point. Take phasing out coal, for instance: More than one-third of the entire coal consumption of the world, more than two billion tonnes, takes place in China, which is growing at nine percent a year. The central government cannot phase out the use of coal for power generation not only because China has some of the largest remaining reserves of good quality coal in the world, but because it is extracted from no fewer than 28,000 small and unsafe coal mines that take a toll of more than 5,000 lives every year, but which the township and other local governments that run them cannot afford to close because it is one of their main sources of budgetary revenue. As of now, despite more than one effort, no government in Beijing has succeeded in closing down more than a small fraction of these mines.

India, the other very large developing country, also has substantial reserves (approximately 255 billion tonnes) of coal. In order to sustain the seven percent rate of growth it has maintained during the past eighteen years it will have to install another 700,000 megawatts of power generating capacity in the next twenty-four years, i.e by 2031.[17] But although it would dearly like to rely more on nuclear power, till the end of 2008 the path was barred by the refusal of the Nuclear Suppliers' Group to sell it nuclear reactors and other dual use technology.

The same is true of the consumption of oil, where increased efficiency in the industrialised countries is being dwarfed by increased consumption in the developing countries. In 2006 the International Energy Agency estimated that China and India would account for seventy percent of the increase in oil consumption in the next twenty-five years. But it is difficult to make a moral, let alone political, case for forcing them to slow down the growth of their consumption, for in 2006 half of the Chinese people still lived on an average of $460 a year[18], and three-quarters of the Indian lived on even less.

One does not know whether Hansen seriously believes that it is possible to change the laws of any country to prosecute CEOs of oil companies that buy up new energy technologies only to suppress them, or unveil 'revolutionary' ways of economising in the consumption of oil even while their corporations stake out claims to dig for oil in the nature reserves of Alaska and the soon-to-be ice-free Arctic Ocean. But the mere fact that it proved impossible to make Exxon pay more than a pittance for the damage caused by the oil spill from the *Exxon Valdez*, shows that this is a lost cause.[19] Today governments do not govern the transnational corporations; they are governed by them.

Only the carbon tax remains. In the long run it is difficult to see how there can be any meaningful shift to non-fossil fuels without a universally accepted and legislated attempt to make the private cost of using oil or coal reflect its social cost. But somewhat surprisingly, this did not prevent a group of five internationally reknowned economists including two Nobel Laureates from issuing a report under the aegis of the Copenhagen Consensus in 2009 that placed the carbon tax last on their list of effective steps to mitigate the effects of climate change. An interview given to BBC by its Director, Bjorn Lomborg, made it clear that the group had only been thinking of the effect of higher carbon prices on consumption and had not even considered its utility as a means of raising revenue to subsidies a rapid shift to non-fossil fuels.

The all–pervasive pessimism described above is driven by an awareness that the shift out of fossil fuels will be of a completely different kind from the three shifts of society's energy base that have preceded it. These were from human and animal to wind and water power a thousand years ago, from wind and water to coal in the 17th and 18th centuries, and from coal to oil between the end of the 19th and the middle of the

20th century. In these shifts there had been no conflict between the new technologies and the forces of the market because the new sources of energy were more concentrated, more portable and more versatile than the ones they were replacing. The shift therefore led immediately to an increase in productivity and profits, and therefore had no trouble in harnessing the drive and the resources of the private sector.

The shift from fossil to non-fossil fuels that the world is facing is the opposite of the earlier ones, for it requires the world to shift from more to less concentrated, versatile and portable sources of energy. Making them more concentrated and portable requires very large investments. As a result, as of now there are very few alternatives to fossil fuels that hold out the prospect of being profitable to private investors. The shift out of fossil fuels will not therefore take place spontaneously but will have to be forced through against the grain of the market economy. This will require concerted action by national governments to push down their cost and force up those of using fossil fuels to reflect the true social costs of these technologies. Since any such large scale and sustained intervention by the state is bound to be fought by the market.

Indeed the fight has already begun. Within days of 2009 G-8 summit at L'Aquila, in Italy, the American Petroleum Institute sent a secret memo to its members asking them to launch a campaign to frighten people into opposing the US' commitment to cut CO_2 emissions, by organising 'town hall' meetings patterned on those the Obama administration launched to explain its health care plans. The US coal lobby is also immensely influential and few in Washington believe that there will be no further watering down of the targets set in the Waxman–Markey bill on CO_2 emission reductions before it passes the senate.

Finally, spiralling estimates of what it will cost to build zero emission housing in the UK, a requirement postulated by the Gordon-Brown government has the building trade up in arms because it could increase the cost of construction by up to eighty percent.

Even those who believe that it is possible—through appropriate international accords—to provide the necessary incentives to overcome the resistance of the Market, harbour deep misgivings about the speed at which it can be done. The shift in the energy base of the economy from wind and water to coal took almost two centuries to complete and that from coal to oil took approximately seventy years. But if the future of human civilisation is to be safeguarded the shift from fossil fuels—oil and coal—must take place within the next three decades at most. Few believe that this is possible.

Chapter 4

A World in Deadlock

If there has been a turning point in peoples' perceptions it has taken place as a result of the publicity that the issue of global warming has received in the run up to the Copenhagen Climate Change Conference, scheduled for December 2009. Whatever doubts may have lingered over the magnitude of the threat that global warming posed to humanity's future, were swept away when 2,500 scientists from eighty countries assembled at the International Scientific Congress on Climate Change in Copenhagen, in March 2009. The Congress, which was convened to update the evidence on global warming collected by the scientific community, pointed out that even the grim findings of the IPCC, published in its fourth report, were three to four years out of date. The evidence collected since then had reinforced the conclusion that global warming was ceasing to be human-induced, and was becoming self-generating. The oceans were absorbing more of the CO_2 and their acidity was rising far more rapidly than the worst-case estimates that had been presented by the IPCC. The consequences for the algae, plankton and other forms of life in the sea would be catastrophic.

The mean air temperature was also rising faster than had been predicted. Five years earlier scientists had thought that the rise, currently of about one degree Celsius since the beginning of the century, could be contained at around two degrees. But fresh evidence showed that this was more of a wish than a prediction. A three-degree rise was now well on the cards.

The assembled scientists also pointed out that even a two-degree rise in temperature could dry out the Amazon forests sufficiently to make them burst into flame at the slightest provocation. A greater rise could hasten their end. They had also found that the Arctic Ocean ice cap was melting faster than even the most determined pessimists had earlier predicted. The rate at which the Greenland and Arctic ice had melted in summer 2008 was what the IPCC had predicted for 2055!

The following were the key messages of the assembled scientists:

> The worst-case scenarios drawn by the IPCC are being realised. For many key parameters, the climate is already moving beyond the patterns of natural variability within which our society and economy have developed and thrived.
> The evidence being provided by the research community increasingly supports the possibility that the earth will experience 'dangerous climate change'.
> Action has to be drastic, immediate and coordinated. 'Rapid, sustained, and effective *mitigation* based on coordinated global and regional action,' they warn, 'is required to avoid *dangerous climate change, regardless of how it is defined.* Weaker targets for 2020 increase the risk of crossing *tipping points* and make the task of meeting 2050 targets more difficult' (emphasis added).[20]

The signal change between the Copenhagen Congress and earlier climate scientists' meetings was that by the 2009 meeting, the assembled scientists were no longer talking of global warming being a gradual phenomenon. Instead they were giving explicit warnings that Abrupt, re-christened 'dangerous' climate change was no longer 'unlikely' as the IPCC had concluded in its fourth report, but 'very likely'. But they still did not explicitly spell out its implications. They did not, for instance, point out that a three-degree rise in the average atmospheric temperature by the end of the century would translate to a six to nine degree rise in the Arctic and Greenland belt. They did not call a press conference to warn the world that long before the world reached this temperature not only would the remaining summer ice in Arctic sea have melted, but also the entire 3,000-meter-thick Greenland ice sheet. This would raise the sea level not by one but by up to seven meters. But this could easily be the least of our problems by then. The huge amounts of fresh, and therefore low-density, water released by the melting of the Greenland ice cap will lower the density of the surface water in the North Atlantic and prevent it from sinking to the lower depths for its return journey south. This will weaken the Meridional Ocean current in the North Atlantic, the Thermohaline current, far more than the IPCC had anticipated. If paleoclimatic data were any guide pushed far enough, this weakening could trigger another sudden ice age. But long before that happened, huge coastal tracts all over the world would have gone under sea; the coral islands, which house several countries, will have vanished; large tracts in the tropical and temperate zones would have become arid desert, and human beings would be killing each other in the millions to control the temperate lands that remained.

Paul Krugman, the MIT economist and columnist of the *New York Times,* spelled out what the acceleration of global warming could mean for the US:

> The fact is that the planet is changing faster than even pessimists expected: ice caps are shrinking, arid zones spreading, at a terrifying rate. And according to a number of recent studies, catastrophe—a rise in temperature so large as to be almost unthinkable—can no longer be considered a mere possibility. It is, instead, the most likely outcome if we continue along our present course. Thus researchers at M.I.T., who were previously predicting a temperature rise of a little more than 4 degrees by the end of this century, are now predicting a rise of more than 9 degrees.... Temperature increases on the scale predicted by the M.I.T. researchers and others would create huge disruptions in our lives and our economy. As a recent authoritative U.S. government report points out, by the end of this century New Hampshire may well have the climate of North Carolina today, Illinois may have the climate of East Texas, and across the country extreme, deadly heat waves—the kind that traditionally occur only once in a generation—may become annual or biannual events.[21]

Thus even though it too shunned alarmism, the Copenhagen Congress went a good ways towards vindicating James Lovelock's admittedly romanticised concept of the Earth as Gaia—a living, breathing, almost sentient system that, if pushed beyond a point, was capable of taking humanity's future out of its hands.

The Copenhagen Scientific Congress therefore swept away the little pretext that had remained for carrying on with

Business As Usual. But the negotiations that have followed have shown how difficult it is going to prove to convert the growing realisation that the world's governments can no longer gamble upon the scientists being proved wrong, into a concrete plan of action. As a result, barely three months before the decisive Copenhagen Conference on Climate Change—the negotiations are close to a deadlock. Once again close to a deadlock.

The central issue on which it has occurred is should those who are responsible for the build up of emissions be responsible for reducing them or should the burden also be shared by those who are not responsible, but whose current emissions are rising rapidly enough to become a cause for concern in the near future?

The first time this issue arose was during the run up to the framing of the UN convention on Climate Change (UNFCCC) and the Kyoto Climate Change conference in 1997. The developing countries had maintained that since they accounted for only a tiny part of the carbon dioxide released into the atmosphere every year, then the responsibility for bringing emissions down lay mainly with the developed countries. The available data fully support this claim: According to the Carbon Dioxide Information Analysis Centre Rich countries account for about seven out of every 10 tonnes of CO_2 that have been emitted since the start of the industrial era. Historical emissions amount to about 1,100 tonnes of CO_2 per capita for the UK and the US, compared with 66 tonnes per capita for China and 23 tonnes per capita for India.[22] The Kyoto conference accepted this argument. The Protocol therefore bound only the OECD countries (and later Russia) to capping their CO_2 and other GHG emissions at the 1990 level. But the developing country exception proved to be a poison pill. The Clinton administration accepted it and signed the Protocol, but since it was to come into force only in 2005, the Bush administration made it a pretext for withdrawing from the Protocol in 2001. The European countries made a half-hearted attempt to honour their obligations. But with the US, Russia, China, India, and other developing nations making no effort to curb emissions, the protocol did next to nothing to reduce CO_2 emissions. As a result by 2007 US emissions had actually increased by twenty seven percent, and those of the EU by six percent. Even the latter figure was deceptive because the increase in emissions had been slowed down by the collapse of the economies of the 10 east European countries that subsequently became members of the EU.[23] For the world as a whole, energy-related CO_2 emissions rose from 22 billion tonnes in 1990 to 29.9 billion in 2007.

Today no one is seriously pushing the developing country exception any more. For between 1990 and 2008 China's CO_2 emissions have risen from 2.3 billion tonnes to a staggering 6.4 billion tonnes. This is 0.2 billion tonnes more than the emissions of the US in the same year. China became the number one polluter because in 2008 when it commissioned 90,000 MW of coal-fired power generating capacity in a single year. These new plants have added another half a billion tonnes of CO_2 to its emissions. India has also added to global CO_2 emissions rapidly, although from a much smaller base. These went up from 0.6 million tonnes in 1990 to 1.5 billion tonnes in 2007.[24] The alarming rate of increase of CO_2 emissions in China, the prospect of a similar rise in India, and a steady accumulation of evidence that global warming is accelerating dangerously and could lead to 'dangerous' climate change, has finally brought home the fact that there can be no exceptions to the fight against it.

But brought home to whom? If one were to judge by the preparatory rounds of discussions, very little has changed since 1997. Indeed the negotiating stances being adopted by the developed and developing countries have increasingly begun to resemble those of the Uruguay round of trade negotiations that were held in the latc '80s and early '90s under the General Agreement on Tariffs and Trade, and the Doha round held under the World Trade Organisation which were stalled for a full year in 2008. They therefore risk suffering a similar fate: if the negotiations do not break down altogether they could end in commitments that both the developed and the developing countries will honour only in form and not in spirit. The critical difference between international trade and climate change is that the world can live without progress in the first but not the second. This basic difference has still to seep into the consciousness of the negotiators.

Following a template laid down at the Kyoto conference the negotiators have discussed four main issues, namely:

1. Mitigation (the reduction of CO_2 emissions over time);
2. Adaptation (adjustment to the impact of global warming);
3. Finance (funding the shift to a low carbon world); and,
4. Monitoring, Reporting and Verification, or MRV.

The following summary shows how far they are from arriving at any form of consensus.

Emission Reduction Targets

In June 2008, when the G8, accepted the target of a fifty-percent reduction in emissions by 2050, it did not specify a base year for the calculation of the reductions. In the US, the Waxman Markey bill has shifted the base year for the calculation of emission reductions from 1990 to 2005. While the EU and Japan are still trying to adhere to the 1990 baseline this has virtually eliminated the moral sanction of the Kyoto accord. Its original goal, of reducing emissions by 6.4 percent below 1990 by 2000 has, therefore, been quietly buried. The Bill, which was passed by the House of Representatives on 25 June 2009, seeks to bring energy-related CO_2 emissions down by twenty percent below the 2005 level by 2020. This is far more modest than the Kyoto target, for it amounts to a reduction of only 4.6 percent below 1990, and that too twenty years later. Its target of a forty-two-percent reduction by 2030 looks like a pie in the sky, because other than supporting the Carbon cap-and trade system the bill has not said a word about how the reduction is to be achieved. But even if the world were to achieve this target, the concentration of CO_2 in the atmosphere would still continue to rise because the annual emission of CO_2 is now more than 50 percent higher than the earth's capacity to absorb it.

Cap-and-trade

Both the US and the EU are committed to bringing down emissions by adopting what is called the 'Cap-and-Trade' system. Under this system the government sets a cap on CO_2 emissions during a particular period that is further broken down by sector, industry, and enterprise. Enterprises that achieve the reduction are allotted carbon credits which they are free to sell in the market to enterprises that have not been able to earn their own. The cost of purchasing the credits gets added to their total cost and therefore to their price. The system therefore puts pressure on the laggards to change their technology, product mix or management practices in order to survive.

Most developing countries have opposed the cap-and-trade system because it is biased in favour of the CO_2 reductions that can be measured and against those that cannot. It will not permit them to claim credits for additions to the GDP through the use of human and animal power such as in handicraft and the unorganised service industries. What is more, as of now, there is no accepted way of measuring the reduction of GHG emissions that occur when farmers change over from traditional, flooded field cultivation of paddy to dry seeding, a change that is rapidly gathering momentum in many parts of India. Nor is there, as of now, an accepted way of measuring the carbon sequestrated in a reforestation programme in different areas and countries.

Martin Feldstein, former Chairman of the US government's Economic Advisory Council President Ronald Reagan's, has pointed out that the system could also harm international trade. This is because it presupposes the existence of a single integrated market, such as the US or EU market. Only then will the carbon credit mechanism put the laggard firms under the pressure to competition to make them change. In a world of sovereign nations, he points out, even if all countries accept the cap-and-trade system and all aim at making the same annual reductions in emission, the impact on costs will be different because of 'differences in initial CO_2 levels and national production characteristics'.[25] The differential impact of the cap-and-trade system on prices in different countries could easily trigger demands for protection. These variations will get further accentuated if different countries interpret their commitments in dissimilar ways, or resort to covert ways of reducing their obligations. Lack of information about what one's competitors in other countries are up to, will make enterprises demand protection from their governments. The world could therefore end up with a spate of additional anti-dumping legislation that would, in turn, further disrupt international trade.

Finance

The US has stated categorically that it is not in any position to provide public funding for any mitigation activities in developing countries. Developing countries should tap the international financial markets for funds and respect intellectual property rights in its adoption of new technology. The EU says there can be some public funding. A part of the funds raised through carbon-credit auctions and the ETS can be placed under the UNFCCC's jurisdiction to fund the above activities. The model would be the UNFCCC's Adaptation fund which mandates that two percent of the money raised from the carbon credits sold under the Clean Development Mechanism should be put in it. The EU, however, is not against developing countries approaching multilateral lending institutions like the World Bank for additional funds for Mitigation.

India's position (which may reflect that of all developing countries) is that the amounts that can be raised from access on carbon credits are wholly inadequate. At present levels of carbon credit sales the Adaptation fund is going to receive no more than $100-150 million a year. Even a five-percent access for the mitigation fund will yield less than $400 million. Bilateral aid has almost dried up and multilateral aid, if taken for CO_2 reduction, will eat into investments intended for other development projects. India has reminded the OECD countries that the UNFCCC is an international treaty and has created legal obligations, the financing of mitigation activities in developing countries being one of them. To tell them now that they should rely mainly on the market for funds or avail themselves of loans from the world bank and other lending by international institutions is to renege on their obligations under the treaty. It therefore insists that there is no alternative to the provision of public funds. Lastly, the disbursal of mitigation funds must be based on transparent rules drawn up by the UNFCCC and disbursed by an agency set up under it that has members from both developed and developing countries. As of June 2009 the deadlock on these issues was total.

Monitoring, Reporting and Verification

The West's insistence on the cap-and-trade mechanism has led directly to a confrontation over the monitoring, reporting and verification (MRV) of emission reductions. The need to, in effect, police the actions of other countries in order to convince industries at home that they are not facing unfair competition is making the US and EU insist that the provision of funds for Mitigation should give the donors the right to monitor not only the funded projects but also all sectors of the economy. Developing countries fear that this is only the thin end of a largish wedge of coercive investigation backed by penalties in trade that will further reduce their sovereign autonomy on economic issues. The Indian position, for example, is that the UNFCCC imposed the obligation to monitor, report, and verify on each country. Donor countries can have the right to verify CO_2 emission reductions, but only in the funded projects and sectors. This part of the negotiations has begun to acquire ugly undertones. No one has used the word yet but 'cross-retaliation' is very much in the air.

The pre-conference negotiations are also remarkable for one important omission: None of the OECD countries has mentioned the matter of carbon tax. In the same essay where he pointed out the dangers that are inherent in the cap-and-trade mechanism, Feldstein also made a powerful plea for a uniform, universal tax on carbon emissions, pointing out that this is the only way to speed up the shift to low-carbon energy without distorting the markets. What he did not mention was that it is also the only way of financing the huge investments that will be needed to complete the shift out of fossil fuels fast enough to meet the Waxman-Markey target of an eighty-three-percent reduction by 2040. The trade in carbon credits in the European Union is currently of the order of $60 billion a year. But a two-percent tax levied on crude oil at the May 2009 price of $60 a barrel will yield $376 billion a year. Over the thirty years the world may have in which to save itself, this tiny increase in price would yield $11 trillion for re-investment, sufficient to capture most of the carbon dioxide that coal-burning power stations are spewing forth today and, in the longer run, meet a large part of the cost of replacing them with nuclear or solar power plants. But ideology dies hard, and neo-liberalism is no exception.

Aquila: the first Breakthrough

Had the deadlock persisted the Copenhagen conference would have been doomed. But better sense prevailed at the G8 summit at Aquila, Italy, in July 2008. In two separate discussions among the leaders—one among the G8, followed the next day by a meeting of the Major Economies Forum on Energy and Climate (MEF)—the sixteen countries that took part made five significant advances.

All of the participating countries (and not only the G8) agreed that the goal of international action on climate change had to be to limit the global average temperature rise to two degrees Celsius. This followed directly from the definition and official acceptance of the possibility of 'dangerous climate change' at the Copenhagen Science Congress. This was the first of two important concessions made by the developing countries that helped to break the deadlock.

The second related to their responsibility for reducing carbon emissions in absolute terms. In the G8 meeting, members made a formal commitment to reduce their own emissions by eighty percent (or more) by 2050. While 2050 was a long way off, this agreement made it necessary for the developed countries to set targets for the intervening years that were consistent with the target for 2050. The EU had already done this, for it had already set a target of twenty to thirty percent for 2020 and was considering raising it to thirty-four percent for the UK and thirty percent for the rest of the EU.

The developing countries responded by accepting that it would not be possible to reach the target of a fifty-percent overall reduction in CO_2 emissions by 2050, if their emissions keep rising. Thus at some point in the near future their emissions would have to peak in absolute terms and start falling. There was no agreement, however, on how soon this should happen.

The most significant advance from the point of view of the developing countries was a tacit acceptance by the G8, of a proposal made by British Prime Minister Gordon Brown in a speech on 26 June to et up a $100 billion fund to finance the shift from carbon technologies in the developing countries. The Major Economies Forum decided to come back to the subject at the G20 meeting in September 2009, after it had been examined by their finance ministers. The MEF also discussed a Mexican proposal to set up a 'Green Fund' to which all countries—except the very poorest—would have to contribute in order to qualify for assistance.

Lastly, at the MEF Summit the developed and developing countries established new commitments to collaborate on the development, dissemination, and transfer of low-carbon technologies, including solar, CCS, low-carbon vehicles and bio-energy.

Chapter 5

Technology Path to the Future

The G8 summit at Aquila, Italy, in July 2008 broke the deadlock that was developing in the climate change negotiations. But it too shied away from discussing the central challenge that the world faces. This is to find an alternative to fossil fuels as the new energy base of human civilisation. All it did was to discuss targets for CO_2 emissions and ways of financing shifts from fossil fuel technologies. But shifts to what? On this they said not a single word. The black hole that lies at the centre of the climate change discussions therefore remained as large as ever.

The key to limiting the rise in temperature to two degrees Celsius, lies in finding alternative technologies to those we use today, which will either replace fossil fuel with renewable energy technologies, or prevent a large part of the carbon we emit from burning fossil fuels from entering the atmosphere.

Alternatives to Carbon

What are the alternative energy sources to fossilised carbon? Can they be tapped quickly enough, even by the developed countries, to reduce CO_2 emissions by eighty percent in the next forty years? Scientists have been exploring an entire cafeteria of alternatives for the last four decades but there is no consensus on the way forward. Most of them maintain a brave face and say that finding alternatives to carbon is not a problem.

'We already have many tools and approaches—economic, technological, behavioural, management—to deal effectively with the climate change challenge' said the 2500 scientists who attended the Copenhagen Congress on Climate Change, in a letter to the Danish Prime minister Anders Fogh Rasmussen. 'But they must be vigorously and widely implemented to achieve the societal transformation required to decarbonise economies'.[26]

But this brave assertion sounds very much like a whistle in the dark, for it begs the crucial question: if there is a plethora of alternative sources of energy, why are so few of them being used today? Why is it that even their limited use has to be coaxed out of investors with heavy subsidies? The answer is that with the exception of nuclear and wind power, very few of the alternatives to fossil fuels have been developed to a point where their commercial viability can be reliably assessed.

In a report prepared titled 'Breaking the Climate Deadlock–Technology for a low carbon future', the Climate Group, headed by Tony Blair, claims virtually the opposite: In a chapter titled 'Signposts for the Future—Ambitious but Achievable Technology Solutions for a Low Carbon Future', the report lists no fewer than 17 technologies in various stages of development and concludes that apart from nuclear power only wind, solar heating and cooling of buildings, hybrid gas-electric automobile engines and heat pumps have been fully commercialised. Seventy percent of the reduction in emissions by 2020, it concludes, will therefore have to come from improved efficiency in the use of fossil fuels.[27]

There is even less knowledge at present of the stages, on the road out of fossil fuels, at which the various technologies being developed today will become commercially viable. All that the signatories to the UNFCCC know as they go into the Copenhagen summit is that forcing the pace of shift to renewable and low-carbon technologies will require funding on a scale and for a time period that has not been mapped out. None of the donor countries therefore have the faintest idea of what they are getting into. None of them knows what technologies their industries will have to adopt to cut emissions at the rate they have committed themselves to, and how much more these will cost. None of them knows how much more they will have to find to induce the developing countries, who will account for seventy percent of the increase in energy consumption between 2010 and 2030, to forego fossil fuels and adopt carbon saving or non-carbon alternatives. And none of them looks forward to finding out how much their competitive position will deteriorate when their companies have to factor in the full cost of carbon sequestration and avoidance while the governments of the developing countries do not, either because they are exempted or because they are receiving compensatory financing. This is the great unknown that lies behind the lack of progress in the climate change talks.

Even a moment's consideration of the way we live, what we eat and how we travel, forces us to recognise that there is only one final energy shift possible for mankind. That is to go back to the sun, to which we owe our existence. In 2003 the world consumed the equivalent of 5,395 billion tonnes of oil, i.e nearly 15 million tonnes of oil a day! Nothing other than the sun itself can come even close to providing such immense quantities of energy. Solar power is the least tapped source of renewable energy. On an average each square metre of the earth's surface alone receives six kwh of energy from the sun every day, equivalent to 21,000 British Thermal Units every day. The energy received every year from the sun by the land areas alone therefore amounts to more than 2,000 times the total commercial energy consumption projected for 2010.[28] The challenge is therefore to capture it in a form that is sufficiently concentrated, and transportable, to meet the present and future needs of the society we have built. *This challenge has been all but met, and at a near-affordable cost. But so fragmented has been the attention that renewable energy technologies have received till now, that this fact remains virtually unknown.*

After a promising start in the seventies, the development of solar thermal technology was arrested for twenty years during the Indian summer of cheap oil. In 2008 there were only six operational solar thermal plants in the world, and another sixteen under construction. Plans for building another sixty-two plants had been announced, mainly in the US and Spain. But all of them are, or will be, heavily subsidised.[29]

Photovoltaic energy is even more expensive because it does not offer the option of storing the sun's heat to extend the working hours of the plant. Thus, although costs are rapidly coming down, neither solar thermal nor solar photovoltaic energy has passed the Market's test of profitability as yet. If solar energy is thought of simply as a source of electricity, neither will make a significant dent in CO_2 emissions in the crucial decade, from 2010 to 2020, which could decide the fate of the planet.

Electricity, however, is not the only form in which the sun's energy can be captured to meet the needs of humanity.

Another way is as the principal input in the manufacture of a much neglected, but extraordinarily versatile and concentrated energy source, methanol.

The Promise, and Strange Neglect, of Methanol

Methanol has three principal uses: as a hydrogen carrier; as a transport fuel; and as a feedstock for the petrochemicals industry. It is second only to natural gas in its versatility, and has the immense advantage of being easy to store and transport. Unlike ethanol, it can be produced from a bewildering array of so-called 'waste' substances, ranging from municipal solid waste, through crop residues to the flue gas emitted by coal, oil or gas-burning power plants. Thus its production in large quantities and its use in all the three above ways, requires no reorganisation of the existing infrastructure of industry and transport.

Where methanol scores most heavily over ethanol is as a transport fuel replacement for hydrocarbons, for it can be produced from the huge crop residues that the world generates every day. It therefore requires no change in the cropping pattern. On the contrary, it is a boon to farmers because it gives great added value to their crops, no matter what these are . . .

There are, broadly, two ways of transforming crop residues and other biomass into methanol: the thermal and the electrochemical. The thermal route involves the gasification of feedstock (burning in a limited supply of air or oxygen) to obtain a mixture of carbon monoxide, carbon dioxide, and hydrogen, called 'synthesis gas', and then recombining them using catalysts to produce methanol in what is called the Fischer-Tropsch synthesis. Synthesis gas is the basic building block of the entire petrochemicals industry, from plastics to fertilisers, but is most easily converted into methanol.

The electrochemical route involves using electricity to generate hydrogen from water through electrolysis and combining it with CO_2 in a synthesis reaction to produce methanol. The oxygen from the electrolysis plant becomes a saleable by-product. The gasification route is appropriate for the conversion of biomass into methanol. The electrochemical route offers a way of capturing the CO_2 in power station flue gases by converting it into a valuable transport fuel or feedstock. But it can also be produced by extracting CO_2 from the atmosphere, thereby directly reducing the concentration of CO_2 in it.

The single greatest anomaly in the global warming debate, therefore, is the way that it has completely ignored the gateway that methanol provides for the harnessing of both solar and biomass energy to create a low-carbon energy system. In the hundreds of studies that have been conducted to find ways of combating global warming that have been published all over the world, there is hardly any mention of methanol. None of the three major reports of the early 2003—the Exxon Mobil energy projections referred to earlier, the Stern report, and the Fourth IPCC report—mentioned methanol. Even more surprising was its complete, although tacit, exclusion from discussions on the way to capture flue gas CO_2 in the report of the Climate Group, in July 2009, on the eve of the Aquila summit.[30] The Exxon-Mobil-report only talked about cellulosic ethanol, whose economics are not known, and commercial production is still more than a decade away. The Climate Group's report relies on the International Energy Agency's data and concludes that most second-generation bio-fuels will be available commercially only by 2030. Neither the Chinese nor the Indian long-term energy projections contain the any mention of the feasibility of converting flue gas CO_2 into methanol. China is the only country with an active methanol production programme for use as a transport fuel. But it is producing it from coal, and could therefore easily find itself using more fossil energy than it is saving. Not surprisingly most projections estimate that renewables will meet less than five percent of the demand for energy in 2030.

The neglect of methanol becomes even harder to understand because it has always been recognised as a better transport fuel than gasoline. Although it has only half the energy content, it has a much higher octane rating of 123 (as distinct from eighty-seven for regular gasoline) and can therefore be used in much more efficient engines with very high compression ratios. Because it also contains its own oxygen it has a higher flame speed. This makes it possible to use it in engines that run at higher revolutions per minute. These characteristics, plus the fact that it does not explode, but simply burns, when it comes in contact with air, make it far safer for use as a motor fuel. Methanol is therefore the prescribed fuel in the Indianapolis 500, one of the three most famous motor races in the world and in many other motor racing events. According to one estimate, the modifications required in the standard automobile engine and fuel supply system to make it run on methanol cost not much more than $100.[31]

When methanol was first suggested as an alternative to gasoline in the '70s, researchers raised several objections to its use on the grounds that, unlike ethanol, methanol is dangerous to human health even when ingested in relatively small quantities. But while methanol may be less safe than ethanol, it is far less harmful—when touched or inhaled—than gasoline. Numerous studies,[32] including one carried out by the Chinese government,[33] have shown that in similar conditions of exposure, such as at petrol stations, methanol is far less toxic than gasoline.

Methanol was therefore one of the alternatives considered by several governments after the first oil shock in the '70s. New Zealand even prepared plans to replace gasoline with methanol obtained by gasifying timber harvested sustainably from its abundant pine forests (*pinus radiata*). But interest in alternative fuels suddenly evaporated when oil prices collapsed in 1986. Methanol, in particular, was sucked into the technology black hole and has still to emerge. Today, like the fountain pen on our desk that we see but do not perceive, we have not even noticed that methanol provides not only the world's best gateway out of fossil fuels but a way of making a host of already developed solar and biomass-based energy technologies immediately profitable.

The Dark Side of 'Business as Usual'

How did methanol get so completely ignored? The personal experience of an MIT scientist, Thomas B. Reed, in the early '70s may provide a clue. In December 1973, just after the first four-fold increase in crude oil prices, Reed and a colleague, R.M. Lerner, published a paper in the noted magazine *Science,* titled, 'Methanol: A Versatile Fuel for Immediate Use', in which they pointed out that blending methanol with gasoline up to fifteen percent required absolutely no change in the settings of a typical motor car engine. It increased fuel economy by five to thirteen percent, and acceleration up to seven percent, despite the fact that methanol contained only half the energy of gasoline. Methanol, they pointed out, was being produced in large quantities from natural gas, but could be obtained in large quantities by gasifying Urban Solid Waste with oxygen to produce synthesis gas. This could be converted to methanol with the catalysts already in use.

The article was summarised in the *New York Times* and created a great deal of interest. This led to a private grant to MIT, where the two scientists were teaching at the time, of $1,000,000 to test the use of methanol as a blending fuel in motor spirits. But just as the project got to the point where tests would be run on 200 privately owned vehicles in Cambridge, Mass., MIT cancelled the project and withdrew the unspent half of the project money.

On '*Woodgas*', the website of the Biomass Energy Foundation that Reed went on to establish in 1997, he would later describe the series of events of 1973 as follows:

> I became interested in the question of alternate fuels in 1973 when my boss asked me to look into 'hydrogen from sunlight' as a possible new energy source. However, hydrogen does not occur in nature and Mother Nature has had 3 billion years of development on renewable fuels through photosynthesis. So I began inquiring about alcohols as fuels....
>
> ...Methanol is by far the simplest and cheapest ($0.15/gal at that time) synthetic fuel to make synthetically. It is used in preference to gasoline at the race track, and I learned in 1973 that student teams had converted cars to methanol and won clean air car races...

A close friend of mine, Dr John Anderson, had just patented a process for converting municipal waste to synthesis gas which could be used to make methanol. I wrote a draft of the article for *Science* magazine proposing that methanol would be the most likely replacement for gasoline as the oil ran out. I gave the article to another friend, Dr R.M. Lerner to read and comment. He liked the article, but suggested that the transition to a methanol fueled future would be easier if it could be blended in increasing amounts into gasoline, say starting with 5 to 10% and gradually developing more capacity as oil became increasingly difficult to find.

I liked this suggestion, so tested a little methanol in gasoline and found it dissolved at all ratios (at room temperature). Being adventurous and having an aging 1969 Toyota, I got a gallon of methanol from the stockroom and added it to my tank to get a 10% mixture. It started fine and I spent the afternoon doing several errands—didn't notice any difference in the driving. After a few more tanks-full of 10% methanol in my gasoline, I noticed that the 'tank to tank' fuel economy was equal to or better than gasoline. As a chemist, I knew that a gallon of gasoline had twice as much energy as a gallon of methanol, so I didn't believe it, (and) decided to make some more scientific tests. I obtained a fuel burette (graduated cylinder) and mounted it in the window of my car and drove over a test course involving a 1 mile hill climb using 0%, 5%, 10%, 15% and 20% methanol in the burette. My mileage went up from 26.8 MPG on gasoline to 28.8 MPG on 15% methanol (a 7% increase), then declined but was still drivable.

Unfortunately the oil and motor industry was less than enthusiastic. A few months later a grant of $1,000,000 was given to the MIT Energy Laboratory. We were informed that the oil companies were the real experts in this field and so the permission to run the tests was revoked after we had spent 6 months and $50,000 developing our test methods.

In 1975, *Science* sent a reporter, Allen L. Hammond, to MIT to investigate the project's cancellation. Hammond's article appeared in the November 1975 issue of the magazine, titled, 'Methanol at MIT: Industry Influence Charged in Project Cancellation', (vol 190, p. 761, November 1975). In it he concluded that the $1,000,000 grant that MIT received had been given in equal parts by Exxon and the Ford Motor Corporation, and that the influence these companies acquired may have led the head and administrators of the Energy Research Laboratory at MIT to cancel the project.

Hammond unearthed no direct evidence of strings attached to the $1-million grant, but he found that the Energy Research laboratory that MIT had just set up was avidly in search of funds; that the funding had been obtained through active campaigning by the Vice-President in charge of research at MIT; that seven of the 24 members of its advisory board were from the auto and oil industries; and that of the two scientists whose vote against the project proved crucial, one was the head of the automobile research section of MIT and the other was actually on loan from Exxon when he pronounced himself, as an expert, against the methanol testing project. Apparently he felt no qualms about a possible conflict of interest.

Thirty five years later, Reed speculated on his website: 'It is interesting—and fruitless—to speculate on how history might have been different if the US had developed an aggressive synthetic fuel program in the 1970s. The threat of alternate production from gas, coal, or biomass would have tempered future rises in oil prices and reduced our funding of the Near East oil sheikhs. It would have prevented the Iran-Iraq war and the US-Iraq wars, and the terrorist attacks, and 9/11, and saved thousands of lives; it would have also saved millions of dollars. And we would have a much better estimate of the cost and means of producing the alternate fuels that we will need as the oil'.

Chapter 6

Harnessing the Sun

Studies carried out by several teams of scientists and engineers in recent years have shown that methanol can be produced more cheaply from biomass than from any other feedstock except natural gas. In a review article assessing the findings of various research groups, P. Galindo Ciffre and O Badr of the Engineering Study Group of Cranfield University in the UK, found that while methanol was being produced from natural gas at around 200 Euro a tonne, it could be produced from biomass at 300-400 Euro a tonne and from CO_2 at 500-600 Euro a tonne.[34] But despite being more expensive, the CO_2 capture route needs to be trodden because it promises to solve the most difficult problem the world faces: to reduce CO_2 emissions rapidly without banning the use of coal to generate power. This is a practical necessity because no matter how much the world may want to see an immediate shift from coal to nuclear or renewable sources of power, it is simply not going to happen. Four out of the six countries with the largest deposits of coal happen also to be the largest and most powerful countries in the world.[35] All have made it clear that they do not intend to forego the use of this God-given source of energy.

The suggestion that the CO_2 should be converted into methanol instead of being stored underground has been made forcefully by the 1994 Nobel Laureate in Chemistry, George Olah. In a path-breaking book published in February 2006, *Beyond Oil and Gas: The Methanol Economy,* Olah and two colleagues, Alain Goeppart, and G.K Surya Prakash, chalked out a grand vision of how humanity could duplicate photosynthesis in a highly accelerated form and suck carbon dioxide out of the air to bring its concentration back to safe levels.

Living biomass breathes in carbon dioxide from the air during the day, uses the sun's energy to break up its own water into hydrogen and oxygen, and combines the hydrogen with the CO_2 to produce more biomass. In the process it exhales oxygen, which replenishes the oxygen lost in combustion and breathed in by animals. Olah proposes duplicating this cycle by combining carbon dioxide in the air with hydrogen obtained from the electrolysis of water. The other product of electrolysis is oxygen, which can be released in the atmosphere, just as trees release it. Thus in Olah's cycle, electricity is used in place of sunlight to break down the water and a chemical 'shift' reaction (already widely in use) is employed to combine the hydrogen with the carbon dioxide.[36]

In the immediate future, Olah's duplication of photosynthesis offers a way to prolong the use of coal for power generation and thereby promises to avert a head-on conflict between the Market and the Environment that could easily spell the end of human civilisation as we know it today. No matter what promises the developed countries make at Copenhagen, breaking free of coal will be very difficult even for them because at market prices, coal and gas are by far the cheapest sources of electricity in the world, and meet fifty six percent of the world's needs.

The market will therefore resist any attempt to phase them out till changes in relative prices bring forth cheaper alternatives. Furthermore, countries like China and India, which have large reserves of coal, also value it as an indispensable source of energy sovereignty.

Converting the CO_2 in flue gases into methanol offers coal and gas burning power plants an attractive alternative, because the sale of methanol will provide a second stream of revenue that will cover a large part—possibly the entire cost—of CO_2 sequestration. Developing countries, in particular, are therefore likely to view it as a way to honour their commitments to reduce CO_2 emissions without sacrificing the pursuit of a higher standard of living.

How rapidly the world adopts carbon capture via methanol will depend on how far it is willing to go to meet the extra cost of making the process economically viable. Every study of this process has shown that its viability depends upon the cost of the electricity needed for the process. A detailed feasibility study carried out by D Mignard, M Sahebzada, J.M. Duthie, and H.W. Whittington in 2002 showed that if surplus power from the off-peak generating capacity of a thermal power station was used to produce hydrogen, methanol could be produced for as little as twenty-two pence a litre.[37]

This calculation was based upon a study of the capture of only a small part of the CO_2 emitted by a 1000-MW thermal power plant, using 100 MW of the surplus generating of the plan and a small amount of regular (meaning, fossil-fuel-based) power from the plant during the remaining eight hours to run the methanol synthesis plant.

But if all of the CO_2 emitted is to be captured, the plant will need vast amounts of additional energy to produce hydrogen. If this is provided by a nuclear power plant, the cost of production of methanol will still be a comfortable thirty to thirty-three pence a unit. But if solar energy is used and paid its current cost of generation of 7.5 pence (11 US cents) a unit, the cost of producing methanol will rise to between fifty and fifty-two pence a unit. Since methanol has half the energy content of gasoline, a straightforward conversion makes its cost in terms of gasoline more than a pound ($1.54 per litre). However such comparisons do not take into account the superior combustion efficiency of methanol, described earlier, which is reflected by the increase in mileage that cars give when gasoline is blended with methanol (or ethanol). But even if it is priced purely on the basis of its energy content, methanol is likely to be competitive with gasoline in Europe if it is exempted from all taxes, so long as the price of crude oil remains at the levels that prevailed in 2009. However, while methanol from biomass may become economically viable in the US if oil prices rise further, methanol from CO_2 will not be able to compete with gasoline until the administration gets up the nerve to impose of a substantial tax on oil-based transport fuels.[38]

Capturing CO_2 by converting it into methanol has an irreducible advantage over storing it underground. While future research and market forces can at best lower the cost of storage, they will without any doubt eliminate the need for subsidies and make the C-to-M technology a paying one. Both solar thermal power and methanol synthesis from CO_2 are sunrise industries whose costs are already falling and will continue to fall with the passage of time Methanol synthesis is a standard chemical process whose capital costs rise by a factor of 1.6 to 1.7 for every doubling of the size of the plant. As all current cost estimates have been derived from pilot to demonstration scale plants, their cost per unit of output will be lower than the ones on which current estimates are based. As for the cost of solar power, it too is scheduled to fall by as much as half during the next decade. This was confirmed by a study of the development of solar thermal power technology by the US National Renewable Energy Technology Laboratory, located in Golden, Colorado. Based on a study of two types of solar power plants, the well-established parabolic trough concentrators, and the newer power tower systems now being constructed at several locations, the NREL has estimated that by 2020 electricity could be produced from power towers for 5.47 cents/kWh and from parabolic troughs for 6.21 cents/kWh. These figures are half or less than the current cost of generation.[39] In other words while CO_2 storage will have to be subsidised so long as there are coal burning power plants, its conversion into methanol is likely to become profitable even at the current low price of crude oil, in no more than a decade. Not only is the financial support required for this energy shift to take place likely to be much smaller, but it will also have to be given for a limited period of time.

Chapter 7

The Limitless Potential of Biomass

As an alternative to fossil fuels, biomass has received scant attention so far. In his seminal work, *Revenge of Gaia* British scientist Lovelock dismissed the possibility that biomass could substitute fossil fuels on the grounds that there is simply not enough feedstock to make a difference. But like almost everyone else, Lovelock was thinking only of ethanol. Once one accepts that methanol is an equally good substitute for gasoline and diesel as a transport fuel, far from being inadequate, the supply of biomass becomes virtually limitless. This is because every kind of biomass waste—from sawdust and black liquor effluents of paper plants, to sewage sludge, plastic, kitchen waste, and crop residues—can be, and has been, gasified at every level, from pilot to near-commercial scale, to produce either electricity or methanol.

There are two abundant sources of biomass that can be utilised immediately to produce methanol: municipal solid and liquid waste and crop residues. Additional biomass can be obtained from reforestation programmes. Indeed, the certainty that there will be an assured market not only for the timber that normally makes up half of the biomass generated by forests, but for the branches and twigs that normally get left behind, can become a powerful incentive for re-afforestation.

Municipal solid waste

According to projections by the United Nations the urban population of the world, which is now billion, is expected to reach five billion by 2030. Cities are producing close to half a billion tonnes of municipal waste a year.[40] By 2030 this is likely to be close to two billion tonnes. Gasification of the municipal waste offers an attractive alternative to land fills because of the acute shortage, and high value, of urban land. Several technologies for doing this are already in commercial use. More and more municipalities all over the world have therefore begun to do this.[41]

Crop Residues

According to the Food and Agriculture Organization (FAO)'s database on agriculture, in 2006 the global output of cereals exceeded 5,872 billion tonnes.[42] Taking the ratio of husk and straw to grain in rice and wheat plants as a measure, the crop residues from cereals alone exceed nine billion tonnes.[43] The residues from six principal oil-producing crops amount to an additional 1.5 billion tonnes. The global sugar industry generates another 800 million tonnes of bagasse.[44] Overall, therefore, the principal agricultural crops grown around the world generate over eleven billion tonnes of residues. While some of this is used for animal feed, a great deal is burnt in boilers, in the fields to clear them for the next crop, or simply left to decay in the open. Thus biomass residues are not just being wasted but the process of waste itself contributes an estimated thirteen percent of the total greenhouse gas emissions of the planet.

If only half of these crop residues are coupled to small solar thermal power plants and gasified in hydrogen-added, oxygen-fed gasifiers to produce synthesis gas they will suffice to produce more than six billion tonnes of methanol, equivalent to more than three billion tonnes of gasoline. This is substantially more than the 2.104 billion tonnes of transport fuels consumed in 2007.[45] Lovelock's dismissal of biomass as an alternative to fossil fuels was therefore premature, to say the least.

Is Methanol-From-Biomass Commercially Viable?

Methanol has been produced as a by-product of charcoal-making for over two centuries as the industrial demand for it grew in the twentieth century it began to be produced by cracking natural gas and coal, and synthesising it with the aid of catalysts in what is known as the Fischer-Tropsch process. In recent decades the yield of methanol has been vastly increased by raising the temperature and pressure in the gasifier, and adding hydrogen, in addition to oxygen, to the gasification process. Dozens of studies carried out over the past four decades have shown that virtually any type of biomass can be gasified to yield synthesis gas, and thence methanol. In high pressure, fluidised bed, gasifiers that use oxygen in place of air, various types of biomass have yielded 0.5 to 1.2 tonnes of methanol per tonne of air dried biomass.[46] The US Department of Energy estimated some years ago that a tonne of standard biomass would yield 186 gallons (550 kgs) of methanol. Other studies in laboratories as far apart as Germany, Sweden, and Hawaii, have shown that this could be the lower end of possible yields, and that if hydrogen is added to the gasification reaction the yield can be raised to 1.1 tonnes of methanol per tonne of biomass.[47]

In their review of four recent feasibility studies, Cifre and Badr concluded that the cost of producing methanol from biomass would range from 300 to 400 euro per tonne. But this assumed the use of off-peak power from a conventional power plant and no attempt to increase the output by adding hydrogen. However, detailed estimates given by Specht and Bandi, and updated by Cifre and Badr, which are based on a fifty tonne-day pilot plant in which they introduce hydrogen to obtain a yield of 1.2 tonnes of methanol per tonne of dry biomass, gives an estimate of 428 euro per tonne of methanol (33.1 centimes per litre). If conventional power is replaced entirely by solar energy costing 11 US cents a unit (about eight centimes) the estimated cost of production rises to 728 euro per tonne (56 centimes/litre)[48] (*See Annexure 2*).

The conclusion here is the same one we arrived at in the case of methanol from CO_2: producing methanol from biomass is also on the margin of viability and will require only a small rise in gasoline prices above those prevailing in late 2009, and an exemption from all taxes, to make it an attractive investment opportunities.

Black Liquor Gasification

Whereas the technology required for producing methanol from CO_2 is well-established and needs only a little more development to link the electrolysis and CO_2 purification plants with the methanol reformation plant, most biomass gasification technologies are still at the pilot plant stage, and need to be tailored to different types of biomass before they can be put to commercial use.[49] But there is one type of biomass for which the technology is fully developed and is being applied on a commercial scale. This is Black Liquor, a highly toxic but very carbon-rich cocktail of alkalis and other chemicals, that is used to extract the carbon from wood before the latter is turned into pulp. Till a few decades ago black liquor used to be discharged into streams even in Europe and the US. Today it is dried into cakes and used to feed the boilers to generate process heat for the paper plant.

In 2003, as part of a European Union ALTENER project designed to suggest ways of shifting twelve percent of transport fuel consumption to bio-fuels by 2010, five firms in Sweden undertook a two-year research project to study the technical and economic feasibility of converting black liquor into methanol for use as a blending fuel with gasoline. The research was carried out with simulation models developed from data gathered in a fifty tonne-day pilot plant. The purpose of the simulation was to determine the economics of modifying a 2,000 tonne/day paper pulp mill to produce methanol as a by-product.[50] The appraisal showed that if the paper mill were to gasify the black liquor cakes in an oxygen-blown gasifier (that uses pure oxygen instead of air), instead of burning it in a conventional boiler, it would generate all the process heat needed to run the paper mill as a by-product of the gasification process. The results obtained from the pilot plant showed that a methanol plant attached to the paper mill, would yield 1.4 to 1.6 tonnes of methanol for every tonne of dried black liquor. If a small electrolysis plant was attached in order to produce both the oxygen that the reaction needed and the hydrogen required for the complete conversion of the carbon in the gas to methanol, the yield would go up to 2.2 tonnes of methanol for every tonne of black liquor. The study showed that without added hydrogen the process would yield an internal rate of return of 35 percent and recover the investment in modernisation and expansion in 3.1 years. The cost of production of two litres of methanol, equivalent in energy terms to a litre of gasoline, was estimated in 2005 to be thirty-two centimes a litre. This was far lower than the ex-pump cost of gasoline in Europe that year.[51] Since then oil prices have almost doubled.

By 2006 two demonstration plants had been set up in Vaxjo and Pitea in Sweden. In 2007 Chemrec, the lead firm in the Swedish project embarked on building a 600 tonne/day black liquor based methanol plant.at the Eascanaba pulp and Paper Mills in Michigan, USA, but the project had to be shelved in July 2009 because of the recession that hit the world in 2008.[52]

In 2006 the global paper pulp industry produced 607 million tonnes of wood pulp and generated an estimated 1.1 billion tonnes of black liquor. When dried, this yielded 170-200 million tonnes of black liquor cakes. If all of this had been fed into gasifiers enriched with both oxygen and hydrogen it would have yielded more than 500 million tonnes of methanol—sufficient to meet the entire octane boosting needs of the petroleum industry, and reduce CO_2 emissions in the transport sector by a seventh in that year.

Chapter 8

Managing the Shift to a New Energy Base

The technologies described above for making a shift from oil to methanol and associated bio-fuels have a number of inestimable advantages over the alternatives—from hydrogen-fuelled cars to photovoltaic houses and office buildings—that are being discussed today. If adopted rapidly enough they can spell the difference between arresting, maybe even reversing, the concentration of CO_2 in the atmosphere and thrashing around wildly till Greenland becomes a temperate paradise.

1. The first advantage is the saving of time: These technologies already exist. The CO_2-to-methanol process puts together a package of technologies that industry is already using for other purposes. Based on surplus power from conventional sources, methanol has been produced in this way since 1922. There is a commercial scale plant in operation today at Coleson Cove, New Brunswick, Canada. Solar power generation, its most important component if we are to aim at very high levels of CO_2 capture, is maturing rapidly and its cost is falling from one year to the next.

A biomass gasification process for producing methanol from 'black liquor' is close to commercial operation. A variety of gasification processes for municipal solid waste are also in use in various parts of the world. Although expensive in themselves because of high energy and gas cleaning costs, these are attractive to urban authorities because they eliminate the need for landfills.

There are a large number of biomass gasifiers in the market that only need to be fine-tuned in order to work with other forms of biomass before being scaled up. The entire process could take as little as five years.

By contrast, the time it takes to take a completely new technology from the laboratory to full commercial use can be two decades or more. The IEA's bio-fuels II report estimated, for instance, that cellulosic ethanol would not become an important transport fuel before 2030.

2. The second advantage is that several of them are already capable of yielding a positive rate of return to investors if spared the plethora of taxes that are levied on fossil fuels for environmental, or revenue raising, reasons. This is beyond serious doubt when methanol is produced from black liquor, and from CO_2 with surplus conventional, or nuclear, power. While it is true that methanol produced using solar energy is still marginal today, it will become highly profitable by 2020 when oil prices go back at $150 range and the cost of solar energy has been halved. Returns of thirty to forty percent per annum and payback periods on investment of as little as three years will, by then, be entirely achievable.
3. Important as these are, it is the third advantage that is by far the most enticing. A shift of the transport fuel system from gasoline and diesel and of the petrochemicals industries

from oil and natural gas, to methanol requires no serious change in the 'mode of production'. No coal miners will lose their jobs, whether in the USA, China Russia or India. The same oil tankers will continue to supply the same oil pumps in the same gas stations. No one will lose his or her livelihood because of the change. In the same way replacing oil and natural gas with methanol as an industrial feedstock will require only a few minor changes to the production process. No plants will become suddenly obsolete and face closure, none of the service roads and rail spurs built to service these plants will cease to be used. No jobs will be even temporarily lost. The economic relationships upon which the social and political structure of today's world is built will remain pretty much as they are today. This means that industry will not feel the need to fight the shift and, in doing so delay the exit from fossil fuels, for none of its massive existing investments in oil and gas exploration, refining, and cracking for use in industry will have to be phased out prematurely.

4. It also poses no threat to the oil industry, which has already grown used to blending ethanol with gasoline. Since oil is the first of the hydrocarbons whose supply is likely to run out, shifting in stages to methanol by increasing the ratio of admixture to seventy, eighty-five, and eventually a hundred percent, will extend the life of the industry and allow it to increase the price of poil gradually. This may lower the potential annual profit, but will ensure that the profits keep coming in for much longer.
5. Since methanol can be produced from crop residues and other waste biomass, it requires no shift in cropping pattern and, above all, no diversion of food crops from feeding

humans and animals to producing transport fuels. Switching from ethanol to methanol will not even hurt the American farmers who had expanded their acreage under corn and soya bean in order to meet the demand from the ethanol industry, for they will more than make up for what they lose on the sale of food crops through the sale of crop residues.

6. Most important of all, in the context of the targets that are likely to be set at Copenhagen in December 2009, producing methanol from CO_2 will allow the coal industry and coal-based power plants to continue producing and using coal without pushing the world over the tipping point. By giving time to plan the shift out of coal-based power generation, it will remove the single most important hurdle for China, India, Poland, East Germany, the US, and a number of other countries, in their quest to politically manage the shift out of coal-dependence.

The End of Guesswork

Once the world has decided on the precise route it will follow in order to shift to a low-carbon economy, it will have a much more precise estimate of the cost of funding for research and development, the sums it will have to raise and disburse, in the form of taxes, tax breaks and subsidies, to equate returns in these new areas with returns elsewhere in the economy, and the period during which new projects will have to be positively encouraged in these ways before changing prices and costs make it unnecessary to continue. Since the shift to a low-carbon economy will no longer require actual or possible sacrifices of development by the developing countries, the attention of the negotiators will shift to making the calculations, and framing the policies needed, to manage the technology shift. Conflict will therefore give way to cooperation.

Ushering in the Methanol Economy

The very first hurdle to the introduction of any new technology is the attempt of existing technologies to fight back and delay, its induction. This is a romanticised description of the struggle that is inherent in process of creative destruction that Joseph Schumpeter described almost a century ago[53], and led Lovelock to characterise the Earth as a near-sentient Gaia. One recent example will suffice to show that victory of the new over the old is not a foregone conclusion.

In August 2007 the Swedish firm, Chemrec A.B., which had pioneered the design of a gasifier for black liquor described earlier, signed an agreement with an American firm, the Newpage Corporation, to set up a black liquor gasification project for the production of methanol. The project location was chosen as the Escanaba pulp and paper plant in Michigan, USA. It was enthusiastically backed by the Michigan Economic Development Corporation and flagged off by the Governor of the state, Jennifer Granholm. But Newpage was forced to withdraw less than two years later, in July 2009. The reported causes of such withdrawal were the rise in materials, and therefore capital costs, in 2007 and 2008 during the commodity speculation boom that preceded the financial collapse, and the crash of methanol prices during the ensuing global recession.

But the key reason for the project's failure was unveiled by one sentence in the Newpage CEO, Rick Willett's explanation of why the company withdrew from the project: 'Unlike Europe, the demand for methanol and dimethyl ether as transportation fuels has not developed in North America. The lack of demand for these products in our country doesn't support the feasibility of the project'.

In plain language, with no federal taxes on motor spirits and therefore nothing to exempt if one used methanol in place of gasoline; with three-quarters of the methanol sold in the market being produced from untaxed natural gas; with even the resulting low price being driven down further by recession; and perhaps most importantly, with a continuing fifty-one cents a gallon subsidy on ethanol, the US was the last place on earth where a new transport fuel technology could take root. But these 'market environment' issues have not even been touched upon in the Waxman-Markey bill. Nor have they been discussed in even one of the pre-Copenhagen negotiations on climate change.

The reason, as explained above, is that these issues can only arise after a technology path to change has been decided. But so far technology has remained a gaping Black Hole into which ideas have disappeared but nothing has come out.

Since there cannot be two prices of one product in a market, the very first requirement for the switch to methanol as a transport fuel is to raise the cost of producing it from natural gas to a level that allows biomass or CO_2-based methanol to survive in the market. This would require raising the tax-inclusive retail price of natural gas and LPG to levels that reflect their energy content instead of keeping it well below (in comparison to the price of gasoline and diesel) in the mistaken belief that the combustion of gas is more environment-friendly than that of gasoline. In fact while the exhaust from burning natural gas may be less harmful to health than that from gasoline and diesel, it is much more harmful than the exhaust obtained by burning methanol, whose combustion yields only CO_2, steam, and easily filterable traces of formaldehyde.

Second, the shift to methanol, and for that matter any other non-fossil fuel, will be greatly speeded up if all countries agree to levy a single, universal, undifferentiated tax on oil, gas and coal. A uniform, global, tax on carbon which is levied at the entry port or the point of extraction, and used to subsidise the construction of alternative energy projects, could bring about a swift transition from fossil fuel-based to nuclear and renewable energy. The carbon tax has been proposed by climatologists, discussed by economists, and shunned by policymakers in the industrialised countries whose aversion to 'big government' remains undiminished despite the global financial meltdown of 2008. The only reference to a carbon tax in the current debate has been in the Waxman-Markey bill, which requires the US to levy 'a carbon tax' on imports from countries that do not accept mandatory reductions in CO_2 emissions. That is precisely the kind of carbon tax the world does not need.

In order to tilt the scales in favour of alternative technologies, a carbon tax does not have to be punitive. Nor does it have to be levied from the outset on coal, a measure that is bound to generate resistance, and therefore delay, in the developing countries. If methanol is to be the gateway to a low-carbon economy the tax should be small and be levied in the first instance only on oil. Since oil is in danger of running out first, this makes good economic sense anyway. With current oil consumption running at more than 30 billion barrels a year, a two-dollar tax per barrel—less than three percent at the current price of oil—will raise $60 billion a year. This will provide more than enough funds to facilitate the shift to the solar energy and biomass-based technologies proposed above within the next decade.

The Limits of the Methanol Economy

CO_2 from fossil fuels accounted for 29.9 billion tonnes, i.e. fifty-eight percent of the total GHG emissions of forty-nine billion tonnes in 2007. These are generated almost entirely by three sectors, namely, transport, industry, and power generation. According to the fourth IPCC report, transport accounted for 13.1 percent of total GHG emissions in 2004 and industry for another 19.4 percent. The bulk of the latter takes place in the petrochemicals industries. In these sectors methanol can replace fossil fuels altogether. Since the technologies for producing methanol from CO_2 and biomass are close to technical maturity and economic viability, these could reduce fossil fuel based emissions by well over half of the fossil fuel induced a third of the total GHG emissions in at most two decades.

To bring it down further, however, the world will have to move out of coal-fired power generation altogether. The obvious, perhaps the only immediate, alternative is nuclear power. The fear of a world full of reactors that could blow up and kill millions of people, and of a planet littered with waste that will remain radioactive for tens of thousands of years, has kept nuclear power in a green-induced limbo for more than three decades now. But the onrushing threat of 'dangerous climate change' posed by the continuing rise of CO_2 concentration in the atmosphere, and a near-impeccable record of safety in nuclear power reactors outside the former Soviet Union, has begun to tilt the scales the other way. The problem of nuclear waste disposal has also been viewed in a new perspective by attempts to compress and store CO_2 underground, a development that has begun in Europe and the US. As Lovelock has pointed out, full carbon capture in this way will require burying a mountain of compressed or frozen CO_2, a mile-high and 12 miles in circumference, every year. In contrast the annual waste generated by an equal amount of nuclear power will amount to 16 metre cube. The problem of nuclear waste disposal does not look quite so forbidding as it did only a short while ago.

Nuclear scientists believe that nuclear fission is, in any case, a transitional source of energy. In three to four decades it will be replaced by nuclear fusion, whose only 'waste' is water. Thus the problem of waste disposal will also eventually disappear. But even if they are too optimistic—even if nuclear fusion never takes off—it is difficult to see what alternative the world has today. To do away with the world needs an alternate source of electricity that is capable of taking up the base load that coal-fired plants take up today. Renewable sources—like wind, hydro, and solar power—can, at present, supply only peak power needs. With a sufficiently large heat storage capacity, solar power plants can, in theory, provide electricity throughout the day. But the cost of generation in such a plant can only be surmised at present. It will be another decade before solar power becomes as cheap as nuclear power, and even longer before it is made available throughout the day at a comparable price. That is time the world no longer has.

Bringing Down the CO_2 Concentration in the Atmosphere

The world has committed itself to bringing down annual CO_2 emissions by half in the hope that this will stabilise the accumulation of CO_2 at 450 parts per million and that that in turn will hold down the rise in temperature by the end of the century to two degrees Celsius. But the projections contained in the fourth report of the IPCC suggest that this might prove to be sufficient. It will certainly not meet Hansen's requirement of bringing the concentration down to 350 ppm.

Is this even possible? The answer is that the methanol economy also provides a way to do so. As economies continue to grow, but coal-fired power stations are phased out in favour of nuclear or solar power, what will the methanol that was previously being obtained from Flue gas CO_2 do for feedstock? The answer given by Olah, is that it will have to be sucked out of the air! Since the technology was first used commercially in the US in 1922. It poses very few problems other than a modest increase in the cost of producing methanol. Ciffre and Badr have estimated that this may increase the cost of production of methanol from CO_2 by 20 percent.[54] When applied on the scale that a global methanol-based transport fuel industry will need, it will make a substantial contribution to reducing the concentration of CO_2 in the air and start to bring down the concentration of CO_2 in the atmosphere. Using the CO_2 from the flue gas of coal-fired power plants to produce methanol, therefore, gives us the added bonus of providing a seamless route from the present into an increasingly sustainable future.

Can the Energy Shift be Made in Time?

Till very recently only a small fringe of pessimists believed that it was already too late to avert abrupt, or dangerous, climate change. Lovelock was one of them. But after the warning issued by the scientists who assembled in Copenhagen in March 2008, their number has grown dramatically. Pointing out that the current CO_2 concentration would, in any case, make the world grow warmer for a long time, the Copenhagen Science Congress warned that it might already be too late to contain the average rise in air temperature to 2 degrees Celsius. Even containing CO_2 emissions at 450 ppm, they said, is going to be problematic. But even if the world has not yet crossed the tipping point, can the investments in low-carbon technologies on the required scale be made in time? Many doubt that it can. They point out that the last energy shift—from coal to oil—took more than sixty years to complete, from the end of the nineteenth century to well past the middle of the twentieth. The earlier shift—from wind and water power to coal—had taken even a good deal longer.

But it might be misleading to predict the future solely on the basis of the past. The resources at the command of global investors are so much greater now than at the end of the nineteenth century, and technological change is so much faster, that three decades could easily suffice to replace the bulk of the fossil fuel being burned today with renewable sources of energy through the natural obsolescence of the current energy infrastructure of the world. Ethanol has, admittedly, turned out to be a blind alley, for it has created more problems than it has resolved. But the wildfire growth of the ethanol industry in the US after President George Bush announced his fifty-one-cent subsidy per gallon on ethanol in January 2006, illustrates just how fast the global capitalist system can react, once it gets the right signals from the market. In the eighteen months following his announcement, new ethanol plants came up at the rate of one every five days. By mid-2007, when the ethanol bubble burst, there were more than a hundred plants in operation and twice that number either being planned or under construction.

Methanol and solar power plants could proliferate with as much, or even greater, speed.

Methanol from CO_2 and biomass can kick-start this change because, other than the removal of current subsidies and preferential tax treatment on Compressed Natural Gas and Liquefied Petroleum Gas, these technologies will require no interventions by the state in the market if the power is provided by a conventional or nuclear power plant, and at most a limited intervention if the power is supplied by a solar thermal plant.

Chapter 9

A Better World: Social Benefits of a Methanol Economy

The preceding discussion of alternatives to coal and oil has confined itself deliberately to assessing their economic viability in a market that is kept as free from state intervention as possible. But decisions that could shape the future of humanity should not be based upon an analysis of 'private' costs and returns alone. When we start tallying the social benefits of a shift from fossil fuels to a renewable-energy, methanol-centred economy the arguments in favour of a rapid shift multiply many-fold.

The most obvious additional benefit from capturing CO_2 through methanol is a cleaner environment. CO_2 capture will not be possible without cleaning the fuel gas to a level that has been achieved in very few countries in the world. By the same token, the exhaust gas from methanol consists of only water (steam) and CO_2, with traces of formaldehyde that can easily be broken down or filtered out. Both the nitrous and sulphurous oxides released by the burning of coal, oil, and gas, will therefore become things of the past.

Conversion of crop residues into methanol will confer an added bonus by reducing the amount of methane released into the atmosphere. Methane is produced mainly by cattle, and by the cultivation of rice in flooded fields. But a good part is also produced when agricultural and kitchen wastes are composted, ploughed back into the land, or simply allowed to decay in the open air. Compared to CO_2, methane's effect is relatively short-lived, but it causes twenty-four times as much global warming. Thus while the use of crop residues to produce transport fuels is carbon-neutral, the replacement of some methane with CO_2 is likely to reduce the overall greenhouse effect. This will be over and above the reduction that will result from the replacement of oil and gas as transport fuels.

For more than a century now, deforestation has been devastating the world's flora and fauna, changing its weather systems and, slowly but surely, reducing the earth's capacity to absorb CO_2. It is also directly responsible for an estimated seventeen percent of all greenhouse gas emissions. A methanol economy could provide the strong financial incentive the world lacks today for protecting denuded forest land to permit its regeneration, and for farmers to bring large acreages under trees. A reforestation drive may not restore the natural forests that have been lost, but it will neutralise the existing deforestation, reduce pressure on the remaining natural forests, increase the tree cover of the earth, and augment its capacity to absorb carbon from the air.

However, substantial as these benefits could be, they will be dwarfed by the economic benefits that will accrue to the economy, in general, and to rural communities in particular. By far the most important is a sharp rise in farmers' incomes from the same crop. For the by-products, which they consider as waste, will acquire immense value. It will be surprising indeed if they fail to capture some of the added value for themselves by charging more. An example from the Indian sugar industry will suffice to show how much farmers stand to benefit.

An Indian sugar mill that crushes 1,000 tonnes a day of cane and produces approximately 120 tonnes a day of sugar, earns approximately $96,000 a day from its sale. But if instead of burning the 600 or so tonnes of leftover bagasse gasify it and produces methanol, the bagasse will yield over 300 tonnes, or 387,000 litres of methanol per day in a simple oxygen-blown process and up to 600,000 litres a day if hydrogen is added. At an ex-plant price of US 25¢ a litre—about one-fourth of the prevailing price of gasoline—the sugar mill will earn an additional $ 97,000 to $150,000 a day from methanol. The enterprise will be able to double the price it pays the farmer for his cane and still increase its profits sharply.

The impact of a methanol-based energy system upon domestic and international economic relations will be the opposite of what it has been in the case of ethanol. In 2007 US farmers' profits doubled but the spiralling price of corn led to food riots in countries as far apart as Mexico and Sierra Leone. In a methanol economy, farmers' incomes can double and profits rise by even more but without disturbing domestic and international food prices. If cotton stalks, for instance, acquire a market value, the US will be able to lower subsidies to its cotton farmers and allow world cotton prices to rise. If rice, wheat, maize, sugar beet, rapeseed, and other crop residues do the same, the EU will also be able to reduce its agricultural subsidies in line with its trade commitments. Farmers in the developing countries will receive a double benefit: a rise in income because of the sale of crop residues and a rise in the export price of their main crop. Cotton farmers will stop committing suicide in central India and millions will emerge from destitution in Africa and Latin America.

Nor will the benefits be confined to agriculture and agro-industries. Additional income means a rise in consumption. Industry will of course benefit; but much of the resulting demand, especially for services, will be met locally. This will increase rural non-farm employment.

Towards an Industrial renaissance: A cleaner, greener world; a narrowing of the urban-rural income gap; new opportunities for investment and employment; a solution to the perennial problem of agricultural subsidies; and an easing of tensions over international trade—so much good news is bound to make the recipients sceptical. But we need only to step back from the immediate present to see that a methanol economy will begin to heal a much deeper wound in human society.

The onset of the industrial revolution, and near simultaneous harnessing of fossil fuels to industry and transport in the second half of the eighteenth century, broke the umbilical cord that had previously joined agriculture to industry and transmitted increases in productivity in one to the other. No previous energy shift had done this. The harnessing of wind and water power in the ninth and tenth centuries led to the creation of surpluses in agriculture that gave rise within a hundred years in northern Europe to the thousand towns of the Hansa. The intra-European trade that ensued brought the fine products of the orient—spices, fine silks and cottons, gold and jewellery—to Europe and further raised living standards in the Hansa. Rapid improvements in the weaving, dyeing, printing, and processing of woollen cloth in northern Italy caused a sheep farming revolution in northern Europe that led to the first enclosure movement in England. The rise in supply of wool gave birth to a weaving industry in Flanders and the low countries—modern-day Holland, Belgium, and northern France—that began to take the market away from northern Italy and forced centres like Florence to improve the quality of weaving and processing. This led to a change in trading patterns wherein northern Europe ceased to export raw wool to Italy and began increasingly to export coarse woollen cloth for further processing.

Similar changes took place in Ming China and Mughal India. Indian cottons and Chinese silks became coveted throughout the western world. By the middle of the eighteenth century a 'putting out' system of manufacture had become established in all the great textiles weaving centres of India. And a class of traders-cum-financiers had come into being, the richest members of whom far outstripped in wealth their Venetian and Florentine counterparts.[55]

The constant feature of all these changes was the close link between agriculture and industry. The link was not always beneficial to both. The First Enclosure movement in England would have devastated the English countryside had the Tudor monarchy not strained every nerve for more than a century to protect the serfs from summary expulsion. This slowed down the shift into the production of wool and enabled England to get the most out of the shift of technology in industry.[56]

This is the link that the switch from wind and water to fossil fuels, and the industrial revolution, broke in the eighteenth century. Over the next 150 years, industry's upstream links with agriculture weakened, while those with the mining industry grew even stronger. Gradually, agriculture was reduced to the status of a service industry, necessary for human survival but not for its enrichment. The share of agriculture in the GDP fell steadily as did the proportion of people living off the land. The gap between rich and poor—not only in terms of real incomes but also in the security of their lives—widened constantly till it forced the poor to band themselves to fight for their rights. As industrial productivity continued to grow and society became hungrier for energy, it became necessary to find and control ever newer sources of fossil fuels and rare minerals. These are the elements out of which colonialism, the unbridled nation state, the bloody wars of the twentieth century, and the global terrorism of the twenty-first, have been forged. Seen from this perspective, the twentieth century was an aberration in human history. Never before had human beings devoted so much of their time, ingenuity and resources to devising more and more efficient ways of killing their fellow humans.

The shift out of fossil fuels—not simply to any alternate form of energy but specifically to solar and biomass energy—will break the destructive link between human industry and scarce minerals. In doing so it will change the 'base' of human society. Changes in the 'superstructure'—in relations between classes of humans and between countries and races—will inevitably follow. The changes described above that a methanol economy will bring about will be a beginning not an end.

Annexure 1

The economics of methanol production from flue gas carbon dioxide, using solar energy.

(The following feasibility study is based upon the calculations of D. Mignard, M. Sahibzada, J.M. Duthie, and H.W. Whittington, Op.Cit.)

The tables below reproduce: 1. the authors' estimates for the cost of producing methanol from flue gas CO_2, using 100 MW of the power plant's own surplus power during 16 off-peak hours; and 2. using 100 MW of solar thermal power generated at a cost of 11 US cents (7.5 pence) a Kwh. In 2. all other costs are presumed to be the same as in 1.

Capital exp	Pounds m.	Annual variable cost	Original	Solar option	Nuclear option
Electrolysis plant	34.7	Renewable energy	6.00	31.00**	14.1
Extra de-NOx capacity	0.054	Fossil fuel power	0.527	0.527	0.527
Extra FGD capacity	1.368	Cooling water	0.701	0.701	0.701

Capital exp	Pounds m.	Annual variable cost	Original	Solar option	Nuclear option
CO_2 recovery plant	4.46	L.P. Steam	1.68	0.00***	0.00
H2 compress & store	8.40	Feed & chemicals	0.224	0.224	0.224
CO_2 compression	2.13	Total annual cost	9.13	31.452	15.552
Reactor	1.17				
Separation	0.48				
Water purification	0.315				
Total fixed. Cost	53.1				
Annual fixed.cost *	12.56				

Total cost of producing methanol with nuclear power (at 5 cents a unit) = 28.11 m. pounds

Total cost of producing methanol with solar power: 44 million pounds

Output of methanol: 64970 tonnes = 83,811,300 litres

Cost per litre:

1. with nuclear power = 33.5 p/litre
2. with solar power = 52 p. per litre.

*For the basis of calculations see endnote—

** Output of methanol = 178 tonnes a day for 365 days.

*** In larger plants it is assumed that the low pressure stream is obtained from waste heat. The above calculations are based upon the assumption that the electrolysis and methanol plants are located very close to the thermal power plant, to share its infrastructure, and most importantly to use the waste heat from the plant to meet the needs of the methanol synthesis and distillation process. If solar power is used, the location of the power plant is likely to be far from the thermal power station. The power will have to be transported through high-voltage DC transmission lines. The additional cost of this distribution system is not included.

Annexure 2

Economics of biomass gasification as calculated by Specht and Bandi, using solar energy to add hydrogen and generate the oxygen needed for the process. (Specht, Bandi, et. al. See Ref 27 below – costs adjusted for 2005 by Cifre and Badr).

1. Capacity of plant = 50 tonnes air-dried biomass/day (14,400 tonnes of dry biomass/yr)
2. Capital cost = 23.93 m. Euro
3. Cost of electricity
   a) In Original study = 3 centimes/ unit
   b) Using solar power = 7.5 centimes/unit
4. Electricity consumed = 6.68 MW /tonne of methanol

Cost of production of methanol 9 original) = 428.26 Euro/tonne
Cost of production of methanol using solar energy = 728 euro/tonne = 56 centimes/litre.

Annexure 3

China's methanol fuel programme: A Report from the *Energy Security Journal* of the Institute for the Analysis of Global Security. Monday, 06 October 2008.

By Gregory Dolan

In 2007, China firmly established itself as the driver of the global methanol industry. The country became the world's largest methanol producer and consumer. China also leads the world in the use of methanol as an alternative transportation fuel, blending nearly one billion gallons of methanol in gasoline. Taxi and bus fleets are running on high methanol blends (M-85 to M-100), and retail pumps sell low level blends (M-15 or less) in many parts of the country. At the same time, China is developing production capacity for dimethyl ether (DME) – using coal-based methanol as a feedstock – for markets as a blendstock with liquid petroleum gas (LPG) used for home heating and cooking and as a diesel substitute for buses.

Chinese Premier Wen Jiabao recently visited a methanol plant in landlocked Ningxia Hui Autonomous Region in northwest China where he urged workers to work hard to build up an industrial base that meets the standards of scientific development. Shenghua Group's Wu Xiuzhang told China's Xinhua news agency that the country's methanol and DME industry is developing by "leaps and bounds" and may lead to overcapacity. China's methanol production capacity has grown from 6.16 million tons in 2003 to an expected 20.6 million tons in 2008, with actual anticipated output of 13 million tons. By 2010, China's methanol production capacity could reach 37.24 million tons. To put this in perspective, total world production capacity excluding China is expected to be roughly 40 million tons this year (about 13.3 billion gallons or 50 billion liters).

DME production has also increased from 32, 200 tonnes in 2002 to 1 million tons in 2007. Wu cautioned, 'The methanol fuel gas (has) not reached the standard for auto use, and there is still a long way to go for DME to substitute LPG. Under such circumstances, we must not ignore the over-capacity issue'. Research and Markets, an international research firm, reports that demands for methanol fuel as M-15 and other downstream products will increase China's consumption of methanol at an annual average growth rate of 16.6% between 2008 and 2012, which will help sustain high methanol prices in the country.

For more than a decade, provincial leaders in coal producing provinces (Xinjian, Shanxi, Shaanxi, Henan, Inner Mongolia, Beijing Shi, Hebei, Anhui, Guangdong and Sichuan) have been developing methanol fuel demonstration programs. These efforts have involved methanol producers, automakers, and academic institutions. In September 2006, eight leaders provided a report to the Chinese President Hu Jintao entitled 'Suggestion on Promoting Methanol Fuels to Replace Gasoline and Diesel Fuel'. President Hu approved this "Suggestion" and directed the powerful National Development and Reform Commission (NDRC) to explore the use of methanol fuels. The NDRC now considers coal-based methanol to be a strategic transportation fuel and has directed the development of national methanol fuel blending standards.

Those provinces in their 'official' trial stage have already adopted their own methanol fuel specification to allow the demonstration of methanol vehicles. Today, methanol sells for roughly RMB$2,500

(US$350) per metric ton, while wholesale gasoline costs nearly three as much at RMB$7,000 (US$965) per metric ton, which encourages the 'unofficial' use of low level methanol fuel blends even in parts of the country that do not have a methanol fuel specification. According to Mark Berggren, Managing Director of Methanol Market Services Asia, 2.71 million metric tons (nearly one billion gallons) of methanol was blended with gasoline in China last year, with an additional 800,000 metric tons of methanol fuel demand growth expected in 2008.

'In the US, corn ethanol may be king, but in China where pure economics matters most, methanol is the dominant alternative fuel,' said Methanol Institute President & CEO John Lynn. 'With millions of miles of experience in using methanol in cars, trucks and buses, China is showing the rest of the world how clean transportation fuels can be made from coal'.

China has over 200 methanol plants, with a total production capacity in 2007 reaching over 14 million metric tons (4.7 billion gallons). Many of these are smaller plants that are closing as larger, world-scale facilities are built. While demand for traditional methanol markets (formaldehyde, acetic acid, MTBE which is a gasoline additive) are increasing at double-digit rates, the tremendous growth in new capacity will make large volumes of methanol for fuel markets available. If just 5% of China's cars used M-85 or M-100 fuel and another 15% use M-15, China would displace 13 million tons of gasoline (4 billion gallons) and significantly reduce its dependence on imported oil.

In 2007, China imported 47% of its oil, and this year the country passed Japan to become the world's second largest oil importing nation only behind the United States. As Prof Ni Weidou of the Tsinghua University-BP Clean Energy Research and Education Center told Interfax-China, "The clock is ticking, and China needs to start adopting alternative fuels now in order to lessen its dependency." Last year, China banned the use of grain for ethanol production to ensure food supplies. China does have the world's third largest coal reserves – 126 billion short tons – behind the United States and Russia. China is now the world's largest coal producer and consumer, and has begun a strategic shift to produce chemicals and fuels from coal rather than oil or natural gas.

The production of methanol from coal gasification is a mature technology. In the United States, Eastman Chemical produces methanol from coal gasification at a plant in Kingsport, Tennessee that was built with support from the U.S. Department of Energy (DoE). Based on this experience, the U.S. DoE estimates that methanol can be produced from coal for as little as 50¢ per gallon. In China, production costs from coal are generally RMB$800-1,200 per metric ton of methanol (US$110-165/metric ton, or 33¢ to 50¢ per gallon). In addition, coke furnaces in China generate 80 billion cubic meters of waste gas each year, enough to produce 40 million metric tons of methanol, and significantly reduce pollution in the coal-producing regions. Coal-bed methane deposits of 30,000-35,000 billion cubic meters in China represent another significant energy resource as well as a potential danger that claims miner's lives each year. Just 1000 cubic meters of coal-bed methanol can produce one metric ton of methanol.

China's automotive industry is already stepping up to meet this challenge. The country's fastest growing independent automaker, Chery Automobile, has recently completed demonstration work on 20 methanol flexible-fuel vehicles – capable of operating on methanol or gasoline – now ready for full-scale production. Shanghai Maple Automobile has announced plans to build 2,000 methanol cars in 2008. Chang'an has introduced the methanol-fueled BenBen car. Greely Automotive has put its Haifeng methanol car into production. Shanghai-based Huapa Automotive has built a number of methanol fueled cars. Shanghai Automotive Industry Corporation, one of the big 3 automakers in China, is developing a number of methanol-fueled cars. In addition, a number of smaller companies are converting large numbers of cars to methanol operation.

In December 2007, a delegation of Methanol Institute (MI) members meet with the officials responsible for establishing methanol fuel standards in China. The standards work is headed by the Shanghai Internal Combustion Engine Research Institute, which has more than 50 years of engine research and development experience. The methanol standard work is part of China's 863 alternative energy initiative. In March 2007, China's Committee on Standardization requested three methanol standards, as follows:

High Proportion – This is a M-70 to M-85 (high methanol content) standard. A draft is currently circulating with a number of experts in China, and is expected to be presented in final form to the NDRC by the end of 2008.

M-15 – Expected to be completed by June 2009. The 15% methanol component also would include an additive/co-solvent.

Convertible Methanol Fuel – Also expected to be completed by June 2009. This is an M-100 standard that would set the specifications for neat methanol to be used as a gasoline blendstock at the M-15 level.

Methanol has a long history as a vehicle fuel in China, but over the last two years the nation has put an intensified focus on methanol. China is now ready to move from the demonstration of methanol fueled vehicles to large-scale commercial deployment. Mr. Rong Junfeng, a senior official with China Petroleum and Chemical Corporation recently told Xinhua News Service, 'The standards will surely facilitate supervision over the current methanol market, and will define the way of methanol development in the future'.

The MI delegation also visited three cities in Shanxi Province, touring bus and taxi fleet depots and fueling facilities. Mr Peng Zhigui, who runs the provincial methanol fuel office in Taiyuan and is a former Vice Governor of Shanxi Province, recently told the *Financial Times*, 'Shanxi is doing the best job in China in promoting the use of methanol as a fuel. Our aim is to solve the problem of China's oil shortage. We are creating a new kind of energy'. Shanxi currently has 3 million metric tons of methanol production capacity, with another 5 million metric tons under construction. With plans to add an additional 20 million metric tons, Shanxi anticipates reaching as much as 28 million metric tons of coal-based methanol production in the next five years.

Shanxi Province includes 11 cities with a total population of 33 million people. The Shanxi Provincial government provides subsidies to taxi and bus fleet operators to convert their vehicles to M-100 operation. There are now 200 buses and 1,000 taxis operating on M-100 throughout Shanxi. The city of Changzhi operates 80 buses on M-100, out of a total fleet of 500 buses. Each bus has a 200-liter fuel tank, which provides a range of 400 kilometers on methanol. There is a diesel pre-heater on the bus to warm the methanol and assist with cold starts. When the MI delegates spoke with one of the bus drivers, he was very enthusiastic about the methanol fuel.

In Jinzhong, 110 taxis operate on M-100, out of a total fleet of 300 taxis. There are plans to convert the entire fleet to M-100 operation. There is a coke-based 40,000 metric ton methanol plant in Jinzhong that provides fuel for the taxi fleet and consumer purchases of M-15 fuel. M-15 was first used in 2003 in four cities, and is now sold in all 11 cities across the province, in stations operated by PetroChina and Sinopec.

On October 13-16, 2008, Shanxi will showcase their methanol program by hosting the 17th International Symposium on Alcohol Fuels being held at Taiyuan University. The organizers anticipate over 200 international delegates representing government, industry and academia. They are also planning a technical tour to a fueling station and a methanol production plant, and will feature a methanol vehicle exposition.

In December 2004, the China Association of Alcohol and Ether Clean Fuels and Automobiles was formed by the China Petroleum and Chemicals Association and the China Association of Automobile Manufacturers. A non-profit organization that advocates for the use of methanol, ethanol and dimethyl ether fuels, the association has over 100 members representing methanol producers, fuel retailers, automakers, academic institutions and governmental agencies. On December 1, 2007, the Methanol Institute signed a Memorandum of Understanding to develop a framework for a cooperative relationship between the two organizations during a ceremony held in Shanghai, China. The agreement calls for a broad exchange of information between the two associations to help expand the market for methanol as a transportation fuel.

The signing ceremony was held during the Chinese Association's Annual Meeting, which drew well over 200 attendees. During the ceremony, MI's John Lynn stated, "While much of the early work on methanol fuel use was conducted in the United States and Europe, China has now picked up the baton and is leading this race. This is not a sprint or a marathon. The finish line will not be reached tomorrow, but it is not a decade away either ... Methanol can answer many of the energy security issues we all face. We are confident that today's agreement brings us an important step closer to the finish line."

Gregory Dolan is Vice President for Communications and Policy at the Methanol Institute, which serves as the trade association for the global methanol industry.

Source: http://www.ensec.org/index.php?option=com_content&view=article&id=148:chinatakesgoldinmethanolfuel&catid=82:asia&Itemid=324

Selected References on Methanol Synthesis from Biomass and CO_2

[1] IEA (International Energy Agency). Renewables for power generation: status and prospects. Paris, France: IEA/OECD, 2003. Available from: www.iea.org, August 2005.

[2] Hiraoka K. Prospects for a hydrogen-solar energy system. Translated from J. Middle Eastern Stud 2001;35(9). Available from: www.mesj. or.jp/mesj_e/english/pub/english/pdf/mv29n022001p71.pdf. Accessed on August 2005.

[3] Duthie JM, Whittington HW. Securing renewable energy supplies through carbon dioxide storage in methanol. In *Proceedings of the* IEEE power engineering summer meeting, Vol. 1, July 21–25, Chicago, USA, 2002. p. 145–50.

[4] Schulien S, Sandstede G, Hahn HW. Hydrogen and carbon dioxide as raw materials for an ecological energy-technology. Int J Hydrogen Energy 1999; 24:299–303.

[5] Wolf BM. Bio-methanol as a fuel available from regional sources. Proceedings of EUROSOLAR conference on promoting global transfer activities for renewable energies, June 8–10 2001, Berlin, Germany. Bonn, Germany: EUROSOLAR-Verlag; 2002.

[6] Specht M, Staiss F, Bandi A, Weimer T. Comparison of the renewable transportation fuels, liquid hydrogen and methanol, with gasoline – energetic and economic aspects. Int J Hydrogen Energy 1997; 23(5):387–96.

[7] EPA (US Environmental Protection Agency). Methanol basics. Technical report EPA 400-F-92-009. Washington DC, USA: EPA, 1994. Available from: www.epa.gov. Accessed on April 2005. [8] EPA (US Environmental Protection Agency). Clean alternative fuels: methanol. Technical report EPA 420-F-00-040. Washington DC, USA: EPA, 2002. Available from: www.epa.gov. Accessed on July 2005.

[9] Hemmings JW, Todd DM. The Starchem methanol process: a low cost route to methanol and power. NJ, USA: Foster Wheeler

International Corporation; 2005. Available from: www.fwc.com/publications/tech_papers/oil_gas/starchem1.cfm. Accessed on July <YEAR>.

[10] Fiedler E, Grossmann G, Kersebohm DB, Weiss G, Witte C. Methanol. 7th ed. Ullmann's encyclopedia of industrial chemistry. Weinheim, Germany: Wiley-VCH Verlag GmbH & Co; 2005.

[11] Spath PL, Dayton DC. Preliminary screening-technical and economic assessment of synthesis gas to fuels and chemicals with emphasis on the potential for biomass-derived syngas. Technical report NREL/TP-510-34929. Colorado, USA: NREL (National Renewable Energy Laboratory), DOE (US Department of Energy). 2003.

[12] Roan V, Betts D, Twining A, Dinh K, Wassink P, Simmons T. An investigation of the feasibility of coal-based methanol for application in transportation fuel cell systems. Report. USA: University of Florida. 2004.

[13] Hugill JA, Overbeek JP, Spoelstra S. A comparison of the ecoefficiency of two production routes for methanol. Report ECN-I-01-003. The Netherlands: ECN (Energy Research Centre of the Netherlands). 2001.

[14] Rauch R. Biomass gasification to produce synthesis gas for fuel cells, liquid fuels and chemicals. Technology Brief. IEA Bioenergy agreement, Task 33: thermal gasification of biomass (2001–2003). 2002. Available from: www.gastechnology.org/webroot/downloads/en/IEA/TechnologybriefSynthesisGas.pdf. Accessed on January 2005.

[15] Synetix. Synetix low pressure methanol (LPM) process. Cleveland, USA: Synetix. Available from: www.synetix.com. Accessed on January 2005.

[16] Williams RH, Larson ED, Katofsky RE, Chen J. Methanol and hydrogen from biomass for transportation. Energy Sustain Dev 1995;1(5):18–34.

[17] Clean coal technology programme. Commercial scale demonstration of the liquid phase methanol (LPMEOH) process. Topical report 11. USA: Clean coal technology programme, US Department of Energy (DOE). Office of Fossil Energy. 1999. Available from: www.fe.doe.-Gov. Accessed on January 2005.

[18] DOE (US Department of Energy). Commercial-scale demonstration of the liquid phase methanol (LPMEOH) process. Report no. DOE/ NETL-2004/1199. Washington DC, USA: Office of Fossil Energy, DOE. 2003.

[19] Specht M, Bandi A. The methanol cycle – sustainable supply of liquid fuels. Stuttgart, Germany: Centre for Solar Energy and Hydrogen Research (ZSW). 1999. Available from: www.zsw-bw.de. Accessed on January 2005.

[20] DOE (US Department of Energy). Assessment of costs and benefits of flexible and alternative fuel use in the US transportation sector. Technical report 5: costs of methanol production from biomass. Report no. DOE/PE-0097P. Washington DC, USA: Office of Policy, Planning and Analysis, DOE. 1990.

[21] Oullette N, Rogner HH, Scott DS. Hydrogen from remote excess hydroelectricity. Part II: hydrogen peroxide or biomethanol. Int J Hydrogen Energy 1995;20(11):873–80.

[22] Specht M, Bandi A, Baumgart F, Murray CN, Gretz J. Synthesis of methanol from biomass/CO_2 resources. In: Eliasson B, Riemer PWF, Wokan A, editors. Proceedings of the 4th international conference on greenhouse gas control technologies, 30 August–2 September. Pergamon: Amsterdam, The Netherlands; 1998. p. 723–8.

MeOH Unit Production Cost (€/tonne) 300–400 500–600

References

[1] IEA (International Energy Agency). Renewables for power generation: status and prospects. Paris, France: IEA/OECD, 2003. Available from: www.iea.org. Accessed on August 2005.

[2] Hiraoka K. Prospects for a hydrogen-solar energy system Translated from. J Middle Eastern Stud 2001;35(9). Available from: www.mesj. or.jp/mesj_e/english/pub/english/pdf/mv29n022001p71.pdf. Accessed on August 2005.

[3] Duthie JM, Whittington HW. Securing renewable energy supplies through carbon dioxide storage in methanol. In: Proceedings of the IEEE power engineering summer meeting, vol. 1, July 21–25, Chicago, USA, 2002. p. 145–50.

[4] Schulien S, Sandstede G, Hahn HW. Hydrogen and carbon dioxide as raw materials for an ecological energy-technology. Int J Hydrogen Energy 1999;24:299–303.

[5] Wolf BM. Bio-methanol as a fuel available from regional sources. Proceedings of EUROSOLAR conference on promoting global transfer activities for renewable energies, June 8–10 2001, Berlin, Germany. Bonn, Germany: EUROSOLAR-Verlag; 2002.

[6] Specht M, Staiss F, Bandi A, Weimer T. Comparison of the renewable transportation fuels, liquid hydrogen and methanol, with gasoline – energetic and economic aspects. Int J Hydrogen Energy 1997;23(5):387–96.

[7] EPA (US Environmental Protection Agency). Methanol basics. Technical report EPA 400-F-92-009. Washington DC, USA: EPA, 1994. Available from: www.epa.gov. Accessed on April 2005.

[8] EPA (US Environmental Protection Agency). Clean alternative fuels: methanol. Technical report EPA 420-F-00-040. Washington DC, USA: EPA, 2002. Available from: www.epa.gov. Accessed on July 2005.

[9] Hemmings JW, Todd DM. The Starchem methanol process: a low cost route to methanol and power. NJ, USA: Foster Wheeler International Corporation; 2005. Available from: www.fwc.com/publications/tech_papers/oil_gas/starchem1.cfm. Accessed on July.

[10] Fiedler E, Grossmann G, Kersebohm DB, Weiss G, Witte C. Methanol. 7th ed. Ullmann's encyclopedia of industrial

chemistry. Weinheim, Germany: Wiley-VCH Verlag GmbH & Co; 2005.

[11] Spath PL, Dayton DC. Preliminary screening-technical and economic assessment of synthesis gas to fuels and chemicals with emphasis on the potential for biomass-derived syngas. Technical report NREL/TP-510-34929. Colorado, USA: NREL (National Renewable Energy Laboratory), DOE (US Department of Energy). 2003.

[12] Roan V, Betts D, Twining A, Dinh K, Wassink P, Simmons T. An investigation of the feasibility of coal-based methanol for application in transportation fuel cell systems. Report. USA: University of Florida. 2004.

[13] Hugill JA, Overbeek JP, Spoelstra S. A comparison of the ecoefficiency of two production routes for methanol. Report ECN-I-01-003. The Netherlands: ECN (Energy Research Centre of the Netherlands). 2001.

[14] Rauch R. Biomass gasification to produce synthesis gas for fuel cells, liquid fuels and chemicals. Technology Brief. IEA Bioenergy agreement, Task 33: thermal gasification of biomass (2001–2003). 2002. Available from: www.gastechnology.org/webroot/downloads/en/ IEA/TechnologybriefSynthesisGas.pdf. Accessed on January 2005.

[15] Synetix. Synetix low pressure methanol (LPM) process. Cleveland, USA: Synetix. Available from: www.synetix.com. Accessed on January 2005.

[16] Williams RH, Larson ED, Katofsky RE, Chen J. Methanol and hydrogen from biomass for transportation. Energy Sustain Dev 1995;1(5):18–34.

[17] Clean coal technology programme. Commercial scale demonstration of the liquid phase methanol (LPMEOH) process. Topical report 11. USA: Clean coal technology programme, US Department of Energy (DOE). Office of Fossil Energy. 1999. Available from: www.fe.doe.-gov. Accessed on January 2005.

[18] DOE (US Department of Energy). Commercial-scale demonstration of the liquid phase methanol (LPMEOH) process. Report no. DOE/NETL-2004/1199. Washington DC, USA: Office of Fossil Energy, DOE. 2003.

[19] Specht M, Bandi A. The methanol cycle – sustainable supply of liquid fuels. Stuttgart, Germany: Centre for Solar Energy and Hydrogen Research (ZSW). 1999. Available from: www.zsw-bw.de. Accessed on January 2005.

[20] DOE (US Department of Energy). Assessment of costs and benefits of flexible and alternative fuel use in the US transportation sector. Technical report 5: costs of methanol production from biomass. Report no. DOE/PE-0097P. Washington DC, USA: Office of Policy, Planning and Analysis, DOE. 1990.

[21] Oullette N, Rogner HH, Scott DS. Hydrogen from remote excess hydroelectricity. Part II: hydrogen peroxide or biomethanol. Int J Hydrogen Energy 1995;20(11):873–80.

[22] Specht M, Bandi A, Baumgart F, Murray CN, Gretz J. Synthesis of methanol from biomass/CO_2 resources. In: Eliasson B, Riemer PWF, Wokan A, editors. Proceedings of the 4th international conference on greenhouse gas control technologies, 30 August–2 September. Pergamon: Amsterdam, The Netherlands; 1998. p. 723–8.

Endnotes

1. Paul Krugman: Betraying the Planet. Op-ed. New York Times. June 26, 2009.
2. Ibid.
3. Guy Turner: European and Global CCS:the current status and future need. *New Energy Finance 2009*. 29, April 2009. http://ccs.newenergyfinance.com/.
4. Joshua Chaffin: Public Wary of Carbon Capture. The Financial Times London. Global economy section.July 29, 2009. See also Numby Strikes Ohio. http://www.drudge.com/news/124811/numby-strikes-ohio.
5. A report prepared by the World Bank in April 2008 concluded that 75 percent of the increase in food prices since 2006 had been caused by the diversion of corn and soya beans into the production of ethanol and bio-diesel, and that only 3 percent could be attributed to higher consumption in China and India. This was suppressed by the Bank on the grounds that it overemphasised the role of diversion to the fuel sector, and did not sufficiently take into account a number of other causes including inclement weather, but this would only weaken its overall conclusion without invalidating it.
6. David Tennenbaum: No snows on Kilimanjaro. The Why Files, 23 February 2001. http://whyfiles.org/shorties/073glacier_melt/2.html. accessed June 29, 2009.

7. James M. Taylor: Nature Study debunks Kilimanjaro Glacier Myth. Environment and Climate News. The Heartland Institute. February 1, 2004. The article quoted Betsy Mason writing in *Nature.*
8. Agence France Presse: Scientists: New Zealand's largest glacier will disappear. Report on April 24, 2008.
9. Time:Jeffrey Kluger: Global Warming Heats Up. Time, March 26, 2006.
10. James Lovelock: *The Revenge of Gaia.*Earth's Climate Crisis and the Fate of Humanity. Basic Books New York 2006.
11. http://www.sciencemag.org/cgi/content/abstract/316/5832/1735.
12. The precise statement made on this subject by the IPCC fourth report is as follows:
    "Based on current model simulations, it is *very likely* that the meridional overturning circulation (MOC) of the Atlantic Ocean will slow down during the 21st century; nevertheless temperatures in the region are projected to increase. It is *very unlikely* that the MOC will undergo a large abrupt transition during the 21st century.
    Longer-term changes in the MOC cannot be assessed with confidence". Op. cit p. 32.
13. (IPCC 4th report Synthesis report. Table SPM1. in Exec.Summary for policy makers. p.7).
14. 4th IPCCsynthesis report p. 45.
15. Unlike Lovelock, Hansen has not been a doomsayer for very long. As recently as in 2000 he published a paper titled '*Global Warming in the 21st century: An Alternative Scenario*', in which he suggested that it was more important to reduce the emissions of other greenhouse gases, than of CO_2, because while the latter was the single largest contributor to GHGs the burning of fossil fuels also released a variety of other substances into the air, collectively called aerosols, that collectively exercised a cooling effect on the earth by reflecting sunlight away and

causing changes in cloud formation. Between 1850 and 1999 this cooling effect had almost exactly balanced the heating effect of the rise in CO_2 concentrations in the atmosphere. Global warming was therefore being done mainly by the other GHGs, and in particular methane. Hansen therefore proposed tackling these first, for instance by changing agricultural practices and the feed of bovines. However, he had warned even then that the moderating effect of aerosols was more or less exhausted because of the drive, on health grounds to screen out particulates and harmful gases like sulphur dioxide. The warming effect of CO_2 would soon overwhelm the cooling effect of aerosols. The Fourth IPCC assessment report suggests that by 2005 this had begun to happen, for aerosols were offsetting only about 70 percent of the warming effect of CO_2. But even this may have been an optimistic assessment because the range of variability around the IPCC's mean estimates was far too great to repose a high degree of confidence in them.

16. Estimated from IEA: World Energy Outlook 2007.
17. India's energy consumption has grown by six percent a year in the past two decades.
18. Estimate of the National Bureau of statistics of China, for the year 2006.
19. On March 24, 1989 the Exxon Valdez spilt 10.8 million gallons of Prudhoe Bay crude oil inot the sea in the Prince Willam sound, Alaska. After four appeals against lower court judgements over eight years, in 2008 Exxon managed to get an intial penalty of $387 million in actual and $5 billion in compensatory damages reduced by the US Supreme Court to $537.5 million in actual and no punitive damages.
20. *The Guardian*, U.K. 12 March 2009.
21. Paul Krugman: Betraying the Planet. Op-Ed page. *The New York Times*. 26 June 2009.
22. Centre For Science and the Environment, New Delhi. Equity Watch, December 5, 2008.

23. Christopher Flavin: Low carbon Energy: A Roadmap. Worldwatch Institute. Washington D.C.Table 2. P. 11.
24. Christopher Flavin: Ibid.
25. Martin Feldstein: Cap-and-Trade= protectionism? *The Business Standard*, New Delhi. June 27, 2009. Copyright Project Monitor. www.projectmonitor.com.
26. *The Guardian*, op. cit. 12 March 2009.
27. The Climate Group: Office of Tony Blair. Breaking the Climate Deadlock: Technology for a Low Carbon Future.Prepared by Shane Tomlinson E3G. Chapter 1/www.theclimategroup.org/assets/resources/Technology_for_a_low_carbon_future_report.pdf.
28. The land area of the planet is 148.3 million sq. km.Total insolation of the land area is thus 889 trillion kwh per day or 324.85 quadrillion kwh per year. The energy projection for 2010 of 511 quadrillion btu per year is taken from the IEA's world Energy and Economic Outlook 2007, chapter 1. I kwh = 3500 btu.
29. On the basis of a plant life of 25 years and a return/ interest on borrowed capital of 7 percent the levellised cost of generation for the 50MW Andasol power plant in Spain, the first commercial plant set up in Europe, was estimated to be 0.16 Euro per unit. This is more than three times the average cost of coal based power generation in India today. However Andasol's estimates are extremely conservative. Solar plants now in operation or being constructed in the US are expected to yield power at 10 to 11 cents a unit. A study by the National Renewable Energy Laboratory at Golden, Colorado, estimated that the cost of producing electricity could halve by 2020 to between 5 and 6 US cents a unit. At that point a subsidy may no longer be necessary.
30. The report devotes a section to second generation biofuels, but its conclusion that these will not be commercially available much before 2030 flies in the face of the International Energy Agency's assessment in 2008 that the thermo-chemical technologies for biomass based fuels has few technical hurdles left and can be

in commercial operation as early as 2012. This suggests that its authors were thinking mainly of cellulosic ethanol. The report also does not talk about capturing carbon by converting it into methanol.The climate group.: Breaking the Climate Deadlock: Technology for a low carbon future. Report prepared by Shane Tomlinson E3G.

31. Institute for the Analysis of Global Security Report on Methanol. See also interview with Nobel Laureate Dr. George Olah. NPR, April 28, 2006. http://www.npr.org/templates/story/story.php?storyId=5369301.
32. The following extract is from a detailed study by Malcolm Pirnie: Evaluation of the Fate and Transport of Methanol in the Environment, prepared for the American Methanol Institute 800 Connecticut Avenue, NW, Suite 620 Washington, DC 20006:

    "Methanol's properties and toxicity are well understood. According to the extensive literature reviewed for this study, methanol is neither mutagenic nor carcinogenic. Human exposure to methanol can occur via the inhalation, ingestion, or dermal contact pathways. Inhalation of methanol vapors can cause irritation of the mucus membranes, dizziness, nausea, headaches, and blurred vision if exposure at high levels occurs. While inhalation is the most common route of exposure to the body, ingestion represents the most serious acute health hazard due to the much higher volume of methanol which can be ingested relative to the volume which can be inhaled. The effects of ingestion follow the same pattern described for inhalation. With respect to dermal contact, methanol readily absorbs into the dermal layer with repeated exposure causing eczema, redness, and scaling. However, the current evidence shows that acute toxic effects on humans and some animals from methanol only occur at high doses (>10 mg/l). The U.S. Department of Energy considers gasoline to be "overall" more hazardous to human health than neat (i.e., pure) methanol". (Executive Summary p. ES 6).
33. See appendix on China's methanol programme.

34. P. Galindo Cifre, O. Badr: Renewable hydrogen utilisation for the production of methanol. Energy Conservation and Management. Vol 48. 2007. pp. 519-527.
35. These are the US with deposits of 240 billion tonnes; Russia with 155 billion tonnes; China with 110 billion and India with 55 billion tonnes. Energy Information Administration of the US. World Energy Outlook 2007.
36. George A. Olah, Alan Goeppert and G.K.Suryaprakash: Beyond Oil and Gas: The Methanol Economy. Wiley-VCH. Verlag GmbH & Co. KGaA. Weinheim. 2006. P 243-245.
37. D. Mignard; M. Sahibzada, J.M. Duthie, H.W. Whittington: Methanol synthesis from Flue-gas CO_2 and renewable electricity: a feasibility study. The authors calculated the cost of production for four alternative technologies. Details given pertain to the most economical option. The researchers made their cost calculations on the following assumptions. 100 MW of electricity was provided for 16 hours from the off-peak surplus of a 1000 MW thermal plant at 1.5 pence per unit. Annual fixed cost was calculated on the basis of a plant life of 15 years; depreciation for tax purposes of 5 percent, a tax rate of 35 percent on profits and a minimum average rate of return of 10 percent per annum, post tax. The consumption of renewable electricity turned out to be 288.7 kwh for one kilo mole (44 kgs) of methanol. About 40 percent of this power was used outside the electrolysis plant, i.e., not to produce hydrogen. This means that 21.27 MWh of power was needed to produce one tonne of hydrogen. More details are given in Annexure-1.
38. The lower figure in the range has been introduced because in larger plants the capital cost per unit of output will rise by about 1.85 times for every doubling of output. Per unit fixed costs will therefore fall.
39. NREL: Assessment of Parabolic Trough and Power Tower Solar Technology Cost and Performance Forecasts Sargent & Lundy

LLC Consulting GroupChicago, Illinois Executive Summary: Assessment of Parabolic Trough and Power Tower Solar Technology Cost and Performance Forecasts National Renewable Energy Laboratory. Golden Colorado.

40. Dong Suocheng, Kurt W. Tong and Wu Yuping Municipal solid waste management in China: using commercial management to solve a growing problem* Science Direct. http://www.sciencedirect.com/science?_ob=ArticleURL&_udi=B6VFT-46PBT2G-3&_user=10&_rdoc=1&_fmt=&_orig=search&_sort=d&_docanchor=&view=c&_searchStrId=1015335611&_rerunOrigin=google&_acct=C000050221&_version=1&_urlVersion=0&_userid=10&md5=5c32b56259ff19c47b328d3ccde0b5f8.
41. See for instance NRG's plasma gasification schemes: http://www.westinghouse-plasma.com/projects/projects_under_development.php.
42. FAOSTAT. http://faostat.fao.org/site/567/DesktopDefault.aspx?PageID=567#ancor.
43. One tonne of rice generates 1.5 tonnes of paddy straw. It also generates 0.3 tonnes of rice husk. Wheat generates no husk.
44. http://www.illovosugar.com/worldofsugar/internationalSugarStats.htm and http://en.wikipedia.org/wiki/Sugarcane.
45. Pew Centre for Global Climate Change. www.pewclimate.org/technology/overview/transport.
46. M. Specht*, A. Bandi*, F. Baumgart*, C. N. Murray**, and J. Gretz*** Synthesis Of Methanol From Biomass/Co2 Resources* Center for Solar Energy and Hydrogen Research (ZSW), Hessbruehlstr. 21 C, D-70565 Stuttgart, Germany;** Environment Institute, Joint Research Centre, 21020 Ispra (VA), Italy*** Hydrogen Association Hamburg, 20457 Hamburg, Germany.
47. Material And Energy Balances For Methanol From Biomass Using Biomass Gasifiers Richard L. Bain January 14, 1992 National Renewable Energy Laboratory 1617 Cole Boulevard Golden, *CO* 80401-3393.

48. In practice, however, such plants are likely to use conventional power for a third of the time when solar power (even from a plant with extensive heat storage facilities) is not available, The cost of methanol using both sources of energy is likely to be around forty-nine centimes a litre.
49. Alternatively the biomass can be standardised. A good deal of progress has been made in doing this.
50. Tomas Ekbom, Mats Lindblom, Niklas Berglin and Peter Ahlvik: Technical and Commercial Feasibility: Study of Black Liquor Gasification with Methanol/DME Production as Motor Fuels for Automotive Uses – *BLGMF.* December 2003. EU Altener project. Contract No. 4.1030/Z/01-087/2001.
51. T. Ekbohm: *High Efficient motor fuel production via black Liquor gasification*. Contact tomas.ekbom@nykomb.se.
52. Earth Times July 17, 2009: Newpage to discontinue building bio-fuels plant with Chemrec.A.B of Sweden.
53. See Thomas McCraw: *Prophet of Innovation: Joseph Schumpeter and Creative Destruction*. 2007.
54. P. Galindo Cifre, O. Badr: *Renewable hydrogen utilization for the production of methanol. Energy Conservation and Management.* Vol 48. 2007. pp. 519 -527.
55. Fernand Braudel: *Capitalism and Civilisation.* Volumes 1 and 3. Giovanni Arrighi: *The Long Twentieth Century*. For details of the wealth of Indian traders in the 16th to 18th centuries see Prem Shankar Jha: *Crouching Dragon, Hidden Tiger: Can China and India Challenge the West?* (Counterpoint Press, USA and UK, Penguin Books New Delhi. (Forthcoming.)
56. This was an one of the earliest examples of the need for a strong state to tame the destructive forces inherent in capitalism and harness its immense productive power for the common good. There will be a similar need for state intervention to harness the forces of capitalism and the market for the energy shift that lies ahead.